UNDERSTANDING GROUNDWATER

DATE DUE

JE 7 '06			

Understanding Groundwater

W. Jesse Schwalbaum

NOVA SCIENCE PUBLISHING, INC.
COMMACK, NEW YORK

Art Director: Maria Ester Hawrys
Assistant Director: Elenor Kallberg
Graphics: Susan Boriotti and Frank Grucci
Manuscript Coordinator: Phyllis Gaynor
Book Production: Gavin Aghamore, Joanne Bennette, Michelle Keller
Christine Mathosian and Tammy Sauter
Circulation: Iyatunde Abdullah, Cathy DeGregory and Annette Hellinger

Library of Congress Cataloging-in-Publication Data
available upon request

ISBN 1-56072-404-8

Copyright © 1997 by   Nova Science Publishers, Inc.
6080 Jericho Turnpike, Suite 207
Commack, New York 11725
Tele. 516-499-3103   Fax 516-499-3146
E Mail Novascil@aol.com
Novascience@earthlink.net

Printed in the United States of America

To my fathers

CONTENTS

LIST OF FIGURES

LIST OF TABLES

ACKNOWLEDGMENTS

The assistance of many friends, family and colleagues was essential to the development and preparation of this book. The acknowledgments are roughly chronological.

I would like to heartily thank the following for their assistance, guidance and support in this project, both directly and indirectly: Professor Mohammed Gheith, Professor D. Caldwell, Professor Richard Yuretich, Judith Raiffa, Joan Sevick-Goldstein, Carole Schwalbaum, Brewster Conant, Bernadette Conant, Whitman & Howard, Inc., Jon Beekman, Earth Tech, Professor Steven Mabee, Ben McBride, Cary Parsons and Doug DeNatale.

❶

✿ ✿ ✿ ✿ ✿ ✿

INTRODUCTION

No one who picks up this book needs to be told that our groundwater drinking water supplies, though generally of good quality, are continuously at risk of contamination. Chemical contamination has affected groundwater supplies in every state of the U.S. and in every province and territory of Canada. Several books have been written about the degree of groundwater contamination and the ongoing threats of hazardous wastes and other potential contamination sources. We have read the magazine and newspaper articles and seen the reports on the evening news. Some of us may even be familiar with the specific water quality threats which exist in our own towns and regions.

The fundamental problem has been that, although all of this information and heightened awareness have resulted in attention to the issue (as well as a lot of fear, uncertainty and a distrust of government and industry), all of that attention has not generally resulted in a clearer understanding of the problems associated with groundwater contamination or their potential solutions. However, if we, as a society, are going to effectively address these issues, we will need to become better educated about them. We certainly don't all have to become experts in the science of groundwater, but it would go a long way toward solving groundwater quality problems if more of us knew a little bit more about the water that pours from our kitchen tap: where that water came from, what groundwater really is, how it moves and how it becomes contaminated.

Hydrogeology, the study of groundwater, has historically been viewed as a somewhat mysterious science. Even today, most of the public who rely on this resource know almost nothing about it. But hydrogeology is a science that is becoming increasingly important. It deals with a resource that is necessary to the health and survival of half of the people on this planet (the other half relies on *surface water*, that is, water from reservoirs, rivers and cisterns). As the need for groundwater resources increases steadily and the potential threats to groundwater resources also increase, our society will be faced with some important decisions on how to manage and protect groundwater resources. These are not decisions that can be left completely in the hands of experts and bureaucrats. These are political, economic, life style and health decisions that strike closer to home than most people might think. In order to make these important decisions, many more people will need to consider the mysteries of groundwater. The basic premise of this book is that you don't have to be a rocket scientist, or even a *hydrogeologist*, to understand the basic principles of groundwater.

It has always astonished me how knowledge, even very basic knowledge, about groundwater has remained so obscure to so many people for so long. Groundwater is an extraordinarily vital and commonplace resource. It underlies virtually every square inch of the North American continent (and all the other continents for that matter) and accounts for approximately 50% of the drinking water in the United States. Most people in this country would rate the quality of drinking water as a major environmental concern. And yet, the vast majority of people have only the vaguest notion of where their drinking water comes from. Perhaps the reader knows that her or his water comes from a private well out in the backyard or from a municipal well field. But that particular well was only the water's last stop on a long journey through subsurface sand, gravel or rock fractures. How far back can you trace the water that comes out of your faucet?

There is, of course, a tremendous amount of mythology and misunderstanding about the sources of groundwater. The mythology varies from region to region. In southern New England, where I come from, the prevailing myth is that the groundwater which flows from springs or is pumped from wells made its way down from New Hampshire (or some-times Vermont) in underground rivers. I have heard this myth recounted as far south as Long Island, N.Y. It is an appealing myth. It provides a mechanism for groundwater movement which we can all understand: the flow of groundwater is assumed to be like the flow of water in the rivers which we can see and experience. It also places the source of our water in a cleaner and more pristine environment than the urban, industrial and suburban settings of Massachusetts, Rhode Island, Connecticut or New York. Unfortunately, this myth also provides a false sense of security and may undercut the incentive for groundwater protection.

The keeper of the mysteries in the mythology of groundwater is the dowser, the water witch. The dowser is second in popularity only to the astrologer as the modern inheritor of ancient occult practices. Almost eve-ryone is familiar with the image of a man walking across a field with a forked stick in his hand. Practitioners of dowsing will tell you that it is based on some sort of extrasensory perception, or perhaps, magic. There are no serious scientific investigations which support the efficacy of dowsing. Yet dowsing is far more familiar to the popular mind than any of the basic principles of groundwater science.

The reason that there is more myth than understanding of groundwater is fairly clear: most people have very little, if any, first-hand knowledge of groundwater. Groundwater moves silently beneath our feet on its unseen journey. You can pump groundwater from a well or watch it as it gurgles out of a spring or as it seeps out of a river bank, but by then the water is already out of the ground and can no longer be considered groundwater. Groundwater in its natural habitat is invisible and mysterious.

Since you can never really see groundwater directly, it is an archetypal symbol of the unconscious, the unknown. Vast and hidden, we can only experience groundwater directly when it has left its realm. In this regard, groundwater springs are like dreams. A careful consideration of both can reveal something about their sources.

In the case of groundwater, the information that can be revealed has clearly practical aspects. Knowing more about the occurrence and move-ment of groundwater will enable us to better protect this essential re-source. Many of us express concern about the quality of drinking water and wonder whether the federal or state authority, or local water company or municipality is doing enough to protect our water. Yet how many of us know where the water that we drink comes from? Or the actual quality of that drinking water (ie., what minerals and chemicals are present in the water)? Or even how to find out what the quality is? What are the potential threats to the quality of groundwater that comes out of our kitchen taps? What safeguards are in place to protect groundwater? Can more be done? What can be done to clean groundwater once it has been contaminated?

These are questions that concerned citizens and consumers of groundwater might ask. To some extent, what keeps people from asking them is the veil of mystery around groundwater which we have been discussing. To most of us, our drinking water is simply another utility, like electricity, natural gas or the telephone. We don't have to think about these things very much because there is only a problem if the service is interrupted. But none of these other utilities directly effects our health the way that drinking water does. Our approach to water supply should really be closer to our approach to foods. But there is no labeling system for home drinking water, as there is for most of the foods that we buy.

In spite of the fact that the principles of groundwater remain a mystery to most of the general public, the importance of protecting groundwater resources has been widely recognized for decades. In fact, groundwater protection has become one of the top environmental priorities of the late 20th century. Most major development projects which require an Environmental Impact Report (EIR) or Environmental Impact Statement (EIS) include the potential impacts to groundwater as a major focus of concern. A large percentage of the federal money that is allocated for hazardous waste cleanup efforts across the country is spent on evaluating the impacts to groundwater and remedying those impacts. The single largest environmental threat at the vast majority of federal *Superfund* sites (the most contaminated sites in the United States) is the potential impacts to groundwater resources.

So, although groundwater protection is clearly a vital public health concern, there is relatively little information available to the general public on the subject. Hydrogeology is not a significant part of most public school science curricula, even in earth science courses. You would be hard pressed to find more than one or two books dedicated to the subject in your local public library. In addition, there have been virtually no widely available books on the market which explain this subject to a nontechnical audience. The problem, I think, lies partially in the fact that since groundwater is unseen, there is no image for the public to get a handle on. Advocates of wildlife protection have it much easier. All they need to do in order to raise public interest and money is show a small wide-eyed furry creature. This image creates an emotional response. Show people a gray, fetid river with dead fish floating in it and you will also get a response.

But what can you show people to get them excited about groundwater? Groundwater has relatively few evocative images: a bubbling spring, perhaps, or an old-fashioned, hand-activated pump. Even the most basic concepts of groundwater are a little abstract because they are not within our normal realm of experience. More detailed descriptions of the processes of groundwater flow rely heavily on mathematical equations that would intimidate most people.

Historically, the relative obscurity of groundwater in the public mind has been directly related to its vulnerability to contamination. It was the lack of awareness of groundwater resources that led to the wide scale occurrences of groundwater contamination in the 1970's and 1980's. Since that time, environmental scientists, government regulators and industry have become more aware of the potential threats to groundwater and how to deal with them. But to the general public - the consumers of groundwater, the residents of Love Canal, New York, and Woburn, Massachusetts, and hundreds of other communities which have been affected by groundwater contamination - there is primarily an awareness of the threats. The basic principles of groundwater remain unclear.

There are plenty of highly specialized sciences with enough devotees of their arcane knowledge to put on large conferences, publish monthly journals and perhaps even build institutes - sciences such as systematic theriogenology, helminthology, or Triassic ichthyophagy. To many people, the science of hydrogeology seems to be counted among these enigmatic sciences. I have so often seen that politely puzzled look which follows the response to an inquiry about my line of work.

Hydrogeology. It sounds so familiar in a way. "Hydro" means water, of course. And geology is the study of the earth: rocks, oil and mining, that sort of thing. But what is the connection? At this point I will give my new acquaintance a big hint. It is the study of groundwater. "Ah," they will respond, and then repeat that puzzled look. This word is vaguely familiar also. Ground. Water. But these words together may not evoke a clear image of anything, at least anything someone could make a living at.

The purpose of this book is to introduce the subject of groundwater to those who may rely on groundwater resources or who may be concerned with local groundwater issues but are not necessarily trained in a scientific or technical field. The basic concepts of the occurrence, movement and quality of groundwater are presented in a manner which can be understood by a wide cross section of people.

This book should be of interest to anyone curious about natural or environmental sciences. It should be particularly helpful to: 1) anyone who owns a private well, 2) anyone who obtains drinking water from a public water supply system which includes a well, 3) public and private water purveyors (eg., water companies, water departments or water districts), 4) public officials who make policy related to groundwater (eg., members of water boards, boards of health, conservation commissions, town or city planners, town meeting members), 5) laypersons reviewing environmental impact reports which include groundwater issues, 6) environmental advocates and, 7) earth science teachers and students.

The hydrogeologic principles described in this book apply to virtually all groundwater occurrences. However, in the examples and descriptions of typical groundwater situations which are presented, there is a definite bias toward the types of groundwater conditions which would be encountered in the northeastern United States. As a result, there are specific geologic and hydrogeologic conditions in other parts of the United States and Canada which are not discussed in great detail in this book but which may be important to readers in those regions. This is especially true for readers in arid regions or areas of substantial limestone aquifers. In these cases, I highly recommend that the reader combine the study of this book with a general study of the regional geology and groundwater conditions.

A NOTE ON THE TERM "GROUNDWATER"

According to most dictionaries, the term for underground water can either be "groundwater," "ground-water" or "ground water." Many authors prefer the two word version, "ground water", because it correlates with "surface water," which is always two words. Others prefer the more Germanic style of creating compound words, "groundwater." Both usages are found in the literature. Many years ago I joined the "groundwater" camp without a lot of conviction. Perhaps because groundwater is different enough from surface water to warrant different rules. Besides, why have two words when one word is just as clear and marginally easier to write? The issue deserves no further discussion.

❷

✽ ✽ ✽ ✽ ✽ ✽

Groundwater and Aquifers

If you stand outside in your backyard, or your favorite park, there will almost certainly be groundwater somewhere beneath your feet. What is not certain is just how much groundwater there is and how deep you would have to dig or drill to reach it. Directly beneath your feet will be some variety of what people generally think of as "dirt," but which we will call *soil*. If you were to take a shovel and dig down, you might see that the soil changes color with depth or that it becomes finer or coarser. The soil may start out as sand and become gradually finer until it becomes silt or clay. Or a fine silty soil may abruptly give way to stony gravel at a certain depth. These soils were probably deposited in that location thousands of years ago by running water or glacial ice or, perhaps, wind. On the other hand, the soil may have formed right in place by the slow decomposition of rock over thousands of years.

There is a lot of empty space in this soil. Geologists and soil scientists call this *pore space*. You could try an experiment to see just how much pore space there is in a soil. Fill a glass with some soil from the backyard and fill another identical glass with water. See how much of the water you can pour into the glass full of soil. Be sure to leave enough time for the dirt to get soaked through. If you can pour in about one quarter of the glass of water into the glass of soil, then the soil had a pore space of about 25%. This gives you an idea how much water the soil can hold.

If you were to keep digging downward in your backyard or park, you might dig through hundreds of feet of soil. Or you might strike hard rock within only a few feet. Either way, you will eventually reach *bedrock*. The shovel will do you no good now, it's time to break out the jack hammer. The term "bedrock" refers to a more or less solid mass of rock that has nothing underneath it (for a long, long way) except more rock. Bedrock of many varieties makes up the crust of the earth, which varies in thickness from approximately 6 to 22 miles. Under this crust is the fluid mantle of molten rock which occasionally spews up to the surface from volcanoes. But we won't be digging that deep.

As you dig, you might find that the soil is moist, especially after a recent rainfall. This dampness within the soil is usually referred to as *soil moisture*. The water is held in the soil by a force known as *capillarity*. It is the same force which holds water in a damp towel or sponge. If there is sufficient water in the soil, then gravity will overcome the force of capillarity and the water will seep further downward.

At some point in your excavation, either in the soil or in the rock, you are going to come across real water, not just dampness. Water will begin to seep into the hole you have dug. This is groundwater. The level at which you reach groundwater is usually referred to as the *water table*. If you keep digging further, the water will continue to trickle into the hole you have dug, eventually filling up the hole to the level of the water table.

Figure 2-1 is a schematic profile of a *water table aquifer*. An *aquifer* is a water-bearing body of soil or rock. This figure will give you an idea of the underground territory we have been pretending to dig through. The *unsaturated zone* above the water table is also known as the *zone of aeration* or *vadose zone*. There may be soil moisture here, but the soils are not completely saturated with water. Beneath the unsaturated zone is the *saturated zone*, sometimes referred to as the *zone of saturation* or *phreatic zone*. Just above the water table there is often a *capillary fringe* where water is held above the water table by capillarity, that sponge-like force. Although the capillary zone is saturated with water, the water level within an excavated hole or a well would actually be at the water table (as shown in Figure 2-1). This is because capillary force is able to hold water up in

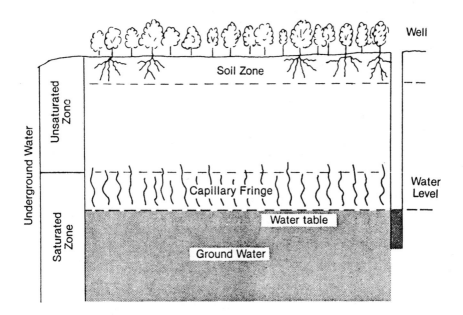

Figure 2-1 Diagram of a Water Table Aquifer
(U.S. Environmental Protection Agency, 1985)

the soil, as in a sponge, but is not able to do this in a larger diameter hole. The difference is in the amount of surface area which water can cling to.

There are some exceptions to the scenario which we have just described, but this is the general case of a water table aquifer for most parts of the country. Groundwater can be found just about anywhere on the earth, it's simply a matter of how deep you have to go to get it and how much there is. That will depend on where you happen to be. If you live in the eastern, midwestern or northwestern parts of the U.S. or Canada, the chances are good that you could drill a reasonably deep well and find enough water for home-use almost anywhere. In areas with drier climates, it is usually more difficult to find an adequate amount of water. Nonetheless, groundwater is relatively abundant. It has been estimated that groundwater makes up approximately 96 to 98% of all fresh water in the

U.S. The remaining 2 to 4% of fresh water is found in lakes, rivers and streams.

A. THE HYDROLOGIC CYCLE

Like the water found in rivers and streams, the water found beneath the ground is constantly moving. It just happens to move very slowly. If you put your ear to the ground, you would not likely hear the gurgling sound of groundwater flowing through the subsurface, even if it were flowing only inches from your ear. Groundwater movement could best be described as a "seeping." Groundwater movement on the order of a foot per day is fairly typical for a sand and gravel aquifer. But although the movement of groundwater is slow, it is steady and its movement is a vital part of what is known as the *hydrologic cycle* - the never-ceasing movement of water between sky, land and sea.

The engine that powers this endless movement is the dynamic interplay between the radiant heat of the sun and the earth's gravity. Energy from the sun causes water to vaporize into a gas. This is called *evaporation*. The warm water vapor rises into the atmosphere where it is subject to the whims of air currents and weather patterns. At some point, the warm moisture-laden air will come into contact with a cold air mass (or land mass for that matter) and the water vapor will condense again into liquid water. In this relatively dense condition, the water droplets will fall to earth under the influence of gravity. Once the water reaches the earth it continues its downward journey over land (in rivers and streams) and under ground (as groundwater) until it reaches a point from which it can no longer flow downward. This will most often be at one of the oceans which cover about three quarters of the earth's surface. Here the water is churned around by the wind and ocean currents, but the only way out again is by evaporation.

The general process I have described above is illustrated in Figure 2-2. It would be helpful to go through the process again in a bit more detail. There are many shortcuts and byways that water can take on its journey from sky to sea and back again. Since the process is cyclical, we could

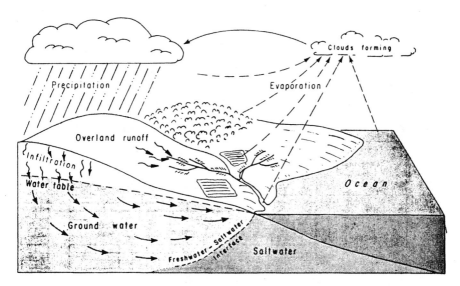

Figure 2-2 The Hydrologic Cycle
(U.S. Environmental Protection Agency, 1985)

start anywhere in Figure 2-2, but it is somewhat customary to begin at the top.

The amount of water which falls to the ground as *precipitation* (a term which includes snow and hail and just about anything that can fall from the sky, including rainfall) can vary widely from place to place and may be seasonal. But whatever the total amount, it is fairly certain that a large percentage of the precipitation will return to the sky relatively quickly, having spent very little time, if any, on the land portion of the hydrologic cycle. Much of the rainfall from a given storm re-evaporates. Some of it re-evaporates before even reaching the ground. A significant amount of water also returns to the atmosphere by means of *transpiration* from plants. Water is drawn up through plants and evaporates from the leaves. For instance, an acre of corn can return 3,000 to 4,000 gallons of water to the atmosphere each day. Since it is difficult to separate evaporation and transpiration, they are often lumped together and referred to as *evapotranspiration*. In the northeast United States, approximately half of all precipitation returns to the atmosphere via evapotranspiration and does not replenish rivers, reservoirs or groundwater reserves. In the arid west, the majority of precipitation will return right away to the atmosphere.

The water that does reach the ground surface will either *infiltrate* into soil or rock crevices or it will wash over the land as *surface runoff* (also called *overland runoff*). Which path the water will take depends on the steepness of the ground surface and its *permeability*. Permeability is a measure of the ability of a material to allow water to flow through it. It is easy to imagine that water is more likely to infiltrate a sandy soil on a flat plain than a compact soil or rock on a hillside. In addition, there is generally a limit as to how much water can infiltrate a given soil in a given period of time. At some point during a rainfall, the *infiltration capacity* of a soil may be exceeded and the excess water becomes surface runoff. You have probably seen this happen during a rainstorm. At the beginning of a storm, the rain is often quickly absorbed by soils, but after a significant downpour the rain begins to puddle and perhaps even flow on the ground surface.

The water which becomes surface runoff will seek the low ground, forming rills and rivulets which drain into brooks and rivers, and eventually flowing out to sea (except in some closed basins in the west, like Death Valley). Along the way, some of the water may take a shortcut back to the atmosphere by evaporating.

The water which infiltrates the soil does not immediately become groundwater. It may be held in the soil above the water table. This soil moisture may eventually evaporate or be picked up by plant roots and transpired back to the atmosphere.

If the rainfall is of sufficient duration, the saturation limit of the soil will be exceeded and water will continue to trickle down through the soil until it reaches the water table. The water table is the surface where soil or rock is saturated with groundwater at atmospheric pressure. The term atmospheric pressure allows us to distinguish the water table from the capillary fringe, which is also saturated. The main point to remember is that the water table is the level at which you would find water in a dug hole or well, as shown in Figure 2-1.

Once water has reached the water table, it is now considered to be part of the groundwater system. From here, it will continue to flow very slowly *downgradient* (the groundwater version of "downhill") until it *discharges* into a spring, river, lake or ocean. Unless, of course, the groundwater is intercepted by a well along the way.

The hydrologic cycle is taking place all around you. The next time it rains try to think about where all that water is going.

B. AQUIFERS

Groundwater flows within an *aquifer* - a water-bearing rock or soil. Not all rocks or soil materials can hold or transport significant quantities of water. The distinction between an aquifer and a non-aquifer can be somewhat subjective. It depends to some degree on what you call "significant quantities of water." If you are looking for a water supply with a well yield of only two or three gallons per minute (gpm) (which would be fine for a one-family home), a slightly fractured mass of granite bedrock might be considered an aquifer. On the other hand, someone looking for a million gallons per day (MGD) for a municipal supply would probably not consider the same rock to be an aquifer at all.

Hydrogeologists generally classify aquifers into two main types: 1) *unconsolidated aquifers* and 2) *bedrock aquifers*. The term unconsolidated refers to soils, as opposed to more or less solid rock. In unconsolidated aquifers the water is present in the pore space between soil grains. This is sometimes referred to as *primary openings* (see Figure 2-3). The water in bedrock aquifers may either be present within primary openings, such as in sandstones, or within *secondary openings*, like rock fractures and solution openings in limestones (Figure 2-3).

UNCONSOLIDATED AQUIFERS

Some of the most productive aquifers in North America are unconsolidated aquifers. Unconsolidated aquifers are composed of soil materials such as sand and gravel. These unconsolidated materials were most often deposited in the geologic past by flowing water of some kind. The same processes are still going on today in river flood plains, coastal plains, del-

Primary Openings

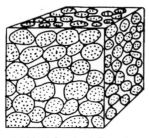

Well-Sorted Sand

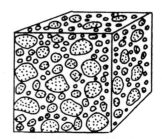

Poorly-Sorted Sand

Secondary Openings

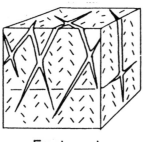

Fractures in
Granite

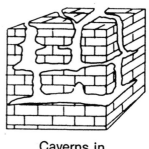

Caverns in
Limestone

Figure 2-3 How Water Occurs in Soil and Rocks
(U.S. Environmental Protection Agency, 1985)

tas, alluvial plains and at the margins of glaciers. You would expect to see
these types of aquifers associated with existing or ancient river valleys and
in areas which have experienced glaciation.

It would be beyond the scope of this book to describe all of the different types of unconsolidated aquifers. However, some of the more common
types of unconsolidated aquifers are: alluvial aquifers, glacial aquifers and
coastal plain aquifers. These are illustrated in Figures 2-4 through 2-6.

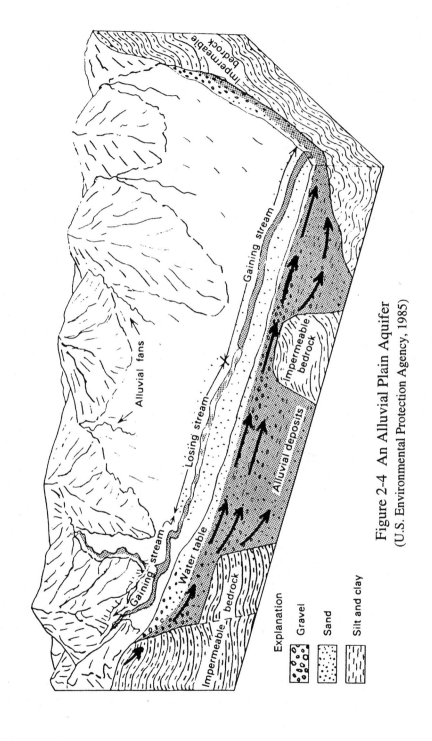

Figure 2-4 An Alluvial Plain Aquifer
(U.S. Environmental Protection Agency, 1985)

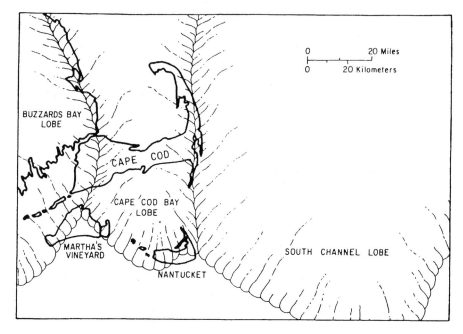

Figure 2-5 The Formation of a Glacial Outwash Aquifer
(Oldale, 1992)

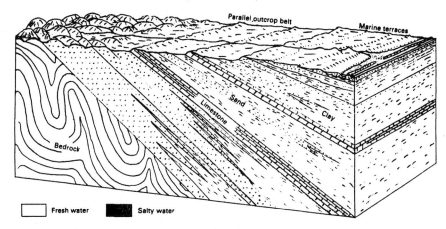

Figure 2-6 A Coastal Plain Aquifer
(U.S. Environmental Protection Agency, 1985)

What most unconsolidated aquifers have in common is that the sands and gravels which form these aquifers were deposited by moving water. Fast flowing rivers can carry tremendous amounts of sand and gravel, but when the waters eventually slow down (such as in a flat valley floor or in a river delta) the coarser, heavier sediments are left behind. Most unconsolidated aquifers are composed of sands and gravels accumulated in this way. The deposits which form alluvial, glacial and coastal plain aquifers are associated, respectively, with rivers, glacial meltwaters and coastal environments.

Most major rivers are associated with an underlying unconsolidated aquifer. The unconsolidated materials which form these aquifers may have been deposited by the river itself, or they may be remnants of an earlier episode of deposition (such as glacial deposition). Soils deposited by a river are referred to as *alluvium* or *alluvial deposits*.

In the Western U.S., alluvial fans and alluvial plains were formed by the erosion of mountain ranges. Swift moving streams carried soils down the steep mountain slopes, depositing them in aprons at the bases of the mountains (see Figure 2-4). The High Plains area is a remnant of a tremendous alluvial plain which extends from South Dakota to Texas and westward to New Mexico. Although these western aquifers are quite large, they are replenished by only scant rainfall.

Much of the Northeast and the Midwest, and virtually all of Canada, bear the remnants of continental glaciation. Glaciers have, on and off, covered a large part of the northern and southern latitudes of the earth over the last million years. Glacial deposits from the most recent episode of continental glaciation are particularly abundant. This episode lasted between 70,000 and 10,000 years ago. In North America this time period is referred to as the Wisconsinan Stage. The best aquifer materials were deposited by meltwater streams during the retreat of glacial ice which began about 20,000 years ago. Examples of regional glacial aquifers include the sandy outwash plains of Long Island in New York and Cape Cod and the islands of Martha's Vineyard and Nantucket in Massachusetts (see Figure 2-5).

The Atlantic and Gulf Coastal Plain consists of a seaward-thickening wedge of unconsolidated deposits extending from Massachusetts southward to Florida, and from Florida westward to Texas. These materials were formed by rivers and streams carrying soils eastward from the Appalachian Mountains to the Atlantic Ocean. The coarser deposits in this deep and extensive geologic feature form local and regional aquifers (see Figure 2-6).

All of these aquifers are similar in that they are composed primarily of varying gradations of sand and gravel, and probably some layers of silt and clay. If you are having trouble imagining what an unconsolidated aquifer looks like, try picturing a large aquarium full of sand. Add water to the aquarium until the sand becomes saturated about half way to the top. You can see where the water line is on the side of the aquarium. This is the water table. Now picture this aquarium as being the size of the county you live in.

BEDROCK AQUIFERS

The groundwater in many bedrock aquifers is found in the cracks, or *fractures,* within the rock. These fractures were created by geologic stresses in the rock (usually some type of folding or faulting) which may have occurred thousands or millions of years ago. The fractures are linear features which range from only inches long to perhaps several miles long.

Digging a hole in your backyard may teach you something about unconsolidated aquifers, but to learn about fractures in bedrock you will have to find an *outcrop* - a surface exposure of bedrock. One good place to find a bedrock outcrop is along a highway or interstate where road cuts provide views of bedrock that are unobstructed by soils and weathering. Look for the dark open cracks of fractures. Sometimes fractures are filled with calcite or quartz crystals and appear light in color. Don't be fooled by the traces of vertical holes which were drilled for blasting. You may even see groundwater seeping from the rock face out of fractures. This phenomena is especially striking in the cold winter months when groundwater seeping from fractures freezes into blue or white mounds of ice.

In order to find groundwater in fractured bedrock it would be helpful to have a method for determining the potential locations of fractures deep within the ground. This is seldom easy to accomplish. Sometimes it is possible to predict the extent and orientations of fractures by examining these features at the surface in bedrock outcrops. It is also possible, occasionally, to determine regional fracture trends by closely examining aerial photographs of a region. These types of examinations are referred to as *fracture trace analyses.* However, it is usually extremely difficult to predict the orientation and water-bearing capacities of bedrock fractures. The fractures may be very local features hidden deep in the rock. Or a prominent fracture may turn out to be filled with silt or clay and therefore not capable of transmitting water.

Most fractured bedrock aquifers produce low to moderate yields of groundwater. There are, however, other types of bedrock aquifers which, though less common in some areas of the country, can produce enormous quantities of groundwater. For instance, some aquifers in limestone contain large openings caused by a dissolving away of some of the rock. These openings may be significantly greater than the pore space in unconsolidated aquifers or the fractures in most bedrock. This accounts for the great yields of wells in some limestone aquifers.

In the Midwest and Western U.S. there are some very extensive sandstone aquifers. These aquifers are similar in many ways to unconsolidated aquifers except for the fact that they are consolidated, that is, the sands have solidified into rock. Even though the sandstone has become "rocklike" there may still be pore space between the grains. Sandstones with a lot of pore space can be very high-yielding aquifers. One example of this is the Dakota Sandstone which is an important groundwater source extending from the Dakotas southward into Texas.

AQUIFERS AND CONFINING UNITS

We have defined aquifers as geologic formations that can convey significant amounts of groundwater and non-aquifers as those which do not. When a non-aquifer, or non-aquifer material, is located adjacent to an aquifer it is called a *confining unit.* Sometimes they are also referred to as *aquicludes* (an awkward term that is not used much these days) or simply as a *boundary.* In general, non-aquifer material is referred to as a confining unit if it occurs above or below the aquifer, and as a boundary when it occurs laterally to the aquifer.

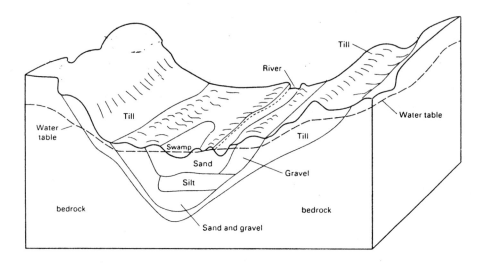

Figure 2-7 Diagram of an Water Table Aquifer
(Haeni, 1995)

Figure 2-7 shows a simple case of an unconsolidated valley aquifer composed of sand and gravel which is surrounded by impermeable bedrock and glacial *till*. In this case, the bedrock and till form a boundary around the valley aquifer. Obviously, the aquifer is not completely bounded because it is open to infiltration of rainwater from above. As mentioned earlier, the level in the ground in which the soils are completely saturated with water is called the water table. Therefore, this kind of aquifer is called a water table aquifer. Note that in spite of the name, the water table is not exactly flat and horizontal. In this case, the water table is sloping down toward the center of the valley.

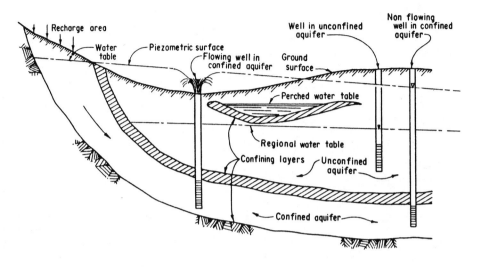

Figure 2-8 Types of Aquifers
(U.S. Department of the Interior)

Figure 2-8 shows a profile of two aquifers separated by a confining layer of clay. The portion of the aquifer which is not overlain by the confining layer is a water table aquifer, but the portion beneath the clay is a *confined aquifer*. There are important differences between water table and confined aquifers in regard to how they respond to a pumping well and also in terms of water supply protection. A water table aquifer is much more vulnerable to contamination from the land surface directly above the aquifer.

Note in Figure 2-8 that the water level (usually indicated by a small triangle pointing downwards) of the well opened to the unconfined aquifer is at the water table but the water level in the well opened to the confined aquifer is higher than the water table. This is a good time to point out that all water levels are essentially measures of pressure, or potential. This is why water levels in a well are sometimes referred to as *piezometric* (from the Greek, "pressure measure") *levels*, or *potentiometric levels*, or *head*. In this example, the pressure (or potential or head) in the confined aquifer is greater than in the unconfined aquifer. This results in an upward potential for groundwater movement from the lower, confined aquifer to the upper aquifer.

If the piezometric level of the lower, confined aquifer were above the ground surface, water would flow out of the well continuously, as shown in the deep well to the left in Figure 2-8. This type of well is sometimes referred to as an *artesian well*, and the aquifer is said to be artesian also. The name comes from the ancient French province of Artesium where water flowed from wells in Roman times (and still does today). Sometimes the term artesian well is used to refer to any well in a confined aquifer. In some regions, particularly New England, the term artesian well is used to denote any well in bedrock. This latter usage has strayed somewhat from the original meaning of the word since bedrock aquifers are not necessarily confined aquifers and the wells rarely flow freely.

Not all confining units are completely impervious to groundwater. A clay layer may be mixed with silts and fine sands, or it may be discontinuous. In this case the material is referred to as a *semi-confining unit* (another term which appears in the textbooks, but is seldom used today, is *aquitard*). A semi-confining unit will allow some groundwater to leak into or out of an aquifer.

Occasionally small lenses of clay or other impervious material can be found in the subsurface. These clay lenses can restrict the downward movement of water and cause a localized puddling of subsurface water which is above the regional water table. This is referred to as *perched groundwater* (see Figure 2-8).

❸

❀ ❀ ❀ ❀ ❀ ❀

The Movement of Groundwater

One of the most important things you will need for a clear under-standing of groundwater is an appreciation of the fact that groundwater is constantly, but very slowly, on the move. In this chapter we will explore some of the principles behind this movement.

It has only been within the past two or three hundred years that theories of groundwater flow have not been dominated by philosophy, church doctrine or mythology. It was Pierre Perrault who, in the late seventeenth century, first demonstrated that groundwater was an integral part of the hydrologic cycle in his book *De l'origines des fountaines* (The Origin of Springs). The Englishman William Smith lays claim to being one of the first hydrogeologists by demonstrating the importance of applying geologic principles to the study of groundwater in the early nineteenth century. The application of mathematics to the description of groundwater flow did not begin until the late nineteenth century. It was at about this time that the scientific discipline we know of as hydrogeology came into being.

Prior to the strict application of the scientific method to inquiries into groundwater, there were many keen observers of groundwater who were adept at locating underground water supplies. Many of these were actually dowsers. Some were engineers or architects. The Roman Vitruvius in the first century made a considerable reputation for himself in this area and his writings indicate a sophisticated understanding of groundwater occurrence. But it took many, many years before these types of observations were put into the context of a rational theory of groundwater flow. Prior to that time, one of the most prominent theories, which dates back to Aristotle, was that groundwater was derived from the sea and that the salt was filtered out as the water seeped landward.

The above theory goes against what many consider to be the first principle of hydrology - that all water flows downhill. Of course, the term "downhill" doesn't apply very well to groundwater since the concept of a "hill" makes no sense below the ground. In more general terms it can be stated that under the force of gravity, all water will seek the lowest possible point. This certainly applies to groundwater and it is the principal concept of groundwater flow.

Let's consider the hydrologic cycle once again from a hydrogeological point of view. When rainfall infiltrates the ground and reaches the water table, it is said to *recharge* the aquifer. When groundwater appears at the ground surface, either in a spring, a lake, a river or the ocean, the water is said to *discharge* from the aquifer. The terms recharge and discharge are used as both verbs and nouns. The water that enters the aquifer is called recharge; that which escapes the aquifer is called discharge. In a general way, all water in an aquifer moves from points of recharge to points of discharge.

Let's go back and consider the aquarium aquifer which we imagined in the previous chapter. The problem with the picture, as we presented it, is that the water in our big aquarium isn't moving. To make this aquarium into a real aquifer we would have to tilt one end of it upward. As we did this, water would begin to flow out the other side. Now, if we continuously trickled (recharged) water onto the top of the sand in the aquarium and let it pour (discharge) out of a hole at the lower end, we would have a dynamic, though extremely simplified, mini-aquifer.

A. THE CONCEPT OF GROUNDWATER BASINS

Many of the concepts used in describing groundwater flow are borrowed from descriptions of surface water flow (the science of *hydrology*, as opposed to *hydrogeology*). The term "downhill" is an obvious example. Another borrowed concept is that of a *basin* or *watershed*. A surface water basin or watershed is determined by the topography of the ground surface and is dominated by a major stream (see Figure 3-1). All of the rain which falls in the basin will eventually drain into the stream (either from direct runoff or groundwater discharge) and that stream will drain either into a larger stream, a lake or, eventually, the ocean.

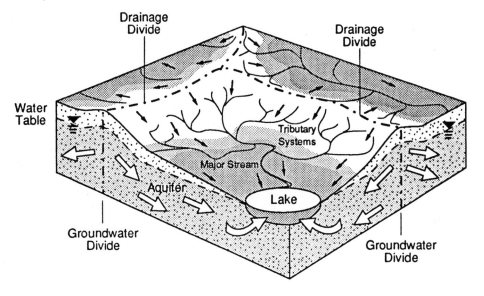

Figure 3-1 A Surface Water Basin or Watershed

In many cases, a valley aquifer will be associated with a surface water basin. In a valley aquifer, groundwater will move from the upland recharge areas to the lowland discharge areas. The area of an aquifer in which all groundwater flows toward a common point or area of discharge, such as a river or lake, is called a *groundwater basin*. The line separating groundwater basins is called a *groundwater divide*. It is quite common for surface water basins and groundwater basins to closely coincide, as does the example in Figure 3-1. However, this is not always the case.

As you might expect, in a valley aquifer the direction of groundwater flow is usually "downhill," that is, down the prevailing topographic slope, as illustrated in Figure 3-2. This is especially true when the aquifer is made up of relatively shallow soils which lie over relatively impervious bedrock. However, there are many possible exceptions to this rule of thumb. Although you can often use surface topography to estimate the groundwater flow direction, this should be considered only a first approximation.

B. Mathematics of Groundwater Flow

The application of mathematics to the description and prediction of groundwater flow was begun in 1856 by Henry Philibert Gaspard Darcy, an engineer who designed the public water system for the city of Dijon, France. Hidden in one of the appendices of his report on the Dijon water system was a discussion of a technique of purifying water by filtration through sand. Darcy developed an empirical equation (that is, an equation which describes observed phenomena rather than one which is derived from a theory) for water flow through sand which is now one of the fundamental equations of groundwater flow and bears his name. Darcy's work was continued and further elaborated by another Frenchman, Arsene Jules Juvenal Dupuit.

By the early twentieth century, most of the important empirical and theoretical mathematics of groundwater flow had been derived. Since that time, groundwater scientists have primarily devoted themselves to devising more efficient techniques for exploring, evaluating and cleaning up aquifers. We will take a closer look at Darcy's equation because it contains within it some of the fundamental principles of groundwater flow.

There are several different ways of expressing Darcy's basic equation. A simple and commonly encountered version is:

$$V = Ki/n$$

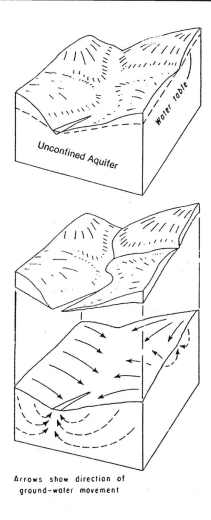

Arrows show direction of
ground-water movement

Figure 3-2 Typical Groundwater Flow in a Valley Aquifer
(Heath, 1987)

Each of the symbols in this equation represents a physical property of groundwater. **V** is the *seepage velocity* of groundwater, **K** is the *hydraulic conductivity*, **i** is the *gradient* of the water table (or *piezometric surface)* and **n** is the *porosity*. These terms will require some explanation.

The *seepage velocity* is essentially the average rate of groundwater flow through a given portion of the aquifer. The expression "average rate" is important here. As soon as we attempt to put a number on groundwater flow, we immediately get into trouble. This is, strange as it may sound, because

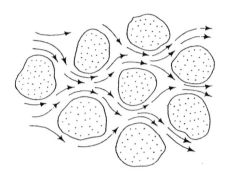

Figure 3-3 Small Scale Movement of Groundwater
(Heath, 1987)

groundwater actually flows differently depending on how closely you look at it. If you could look at an aquifer under a microscope you would see a chaotic jumble of soil grains surrounded by chaotically inter-connected pore spaces, as shown in Figure 3.3. The molecules of water, moving under the influence of gravity, wind their ways through this maze, sometimes moving relatively straight and sometimes taking tortuous routes. For this reason, it is not possible to calculate the flow rates of individual molecules of water. But if you stand back and look at a larger part of a relatively uniform aquifer, the average velocity of water will be a certain rate which we will designate the seepage velocity (V). V is expressed in units of length over time, such as feet per day or meters per second.

Hydraulic conductivity is the measure of the water-transmitting property of a soil. Coarse-grained soils, such as gravel, have high hydraulic conductivities whereas fine-grained soils, like silt and clay, have low hydraulic conductivities. Usually an aquifer is made up of soils with a wide range of different grain sizes which vary somewhat from place to place. Therefore, the hydraulic conductivity of an aquifer will vary, sometimes over a relatively short distance. Usually, to simplify matters, the hydrogeologist or engineer only considers the average hydraulic conductivity over a portion of an aquifer. Often, this value can be only roughly estimated. By this time, you may be beginning to realize that hydrogeology is not as exact a science as nuclear physics.

Hydraulic conductivity, represented by the symbol K, also has units of length over time, just as velocity does. Many people find this confusing at first. However, it is important to realize that hydraulic conductivity is *not* a velocity. If a soil has a hydraulic conductivity of 150 ft./day, that does not mean that groundwater will flow at a rate of 150 ft./day through this material. The groundwater flow rate is dependent on the hydraulic conductivity, the gradient and the porosity.

The *gradient* (i) can be thought of as the slope, or inclination, of the water table. However, confined aquifers do not have water tables, so we need to define gradient more generally. The gradient is the difference in water level (that is, the height to which water will naturally stand in a well) between two points in the direction parallel to groundwater flow, divided by the distance between them. It is similar to the concept of a topographic gradient, or slope. Figure 3-4 illustrates how a groundwater gradient is calculated between two wells. In this example, the gradient is 0.5 ft. / 200 ft. = 0.0025. Note that the gradient does not need to be given in any units (e.g., feet or meters) because it is the same no matter what units of measurement are used.

Although the a gradient can be calculated with data from just two wells, the direction of groundwater flow can only be calculated if the elevation of the water table at three or more points is known. Figure 3-5 gives an example of a site where the water levels at three monitoring wells have been measured. The figure shows a map view of the site of the three wells. It is a relatively simple matter to mathematically triangulate the water levels between these three wells and determine the gradient of the water table

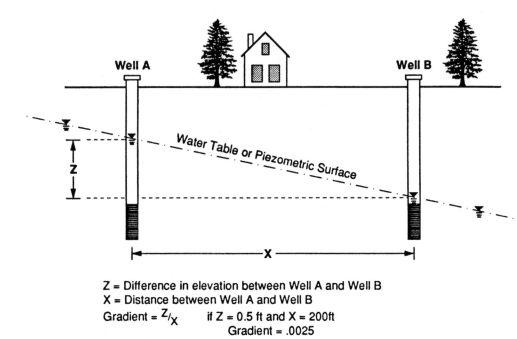

Z = Difference in elevation between Well A and Well B
X = Distance between Well A and Well B
Gradient = Z/X if Z = 0.5 ft and X = 200ft
                              Gradient = .0025

Figure 3-4 Calculating a Groundwater Gradient

surface. Based on these levels, lines of equal groundwater elevation are drawn. These lines are called *contours*, specifically, *water table contours* (or sometimes just groundwater contours). It is exactly the same concept used in developing topographic maps of a land surface.

In the example in Figure 3-5 the gradient can be determined by first drawing a line perpendicular to the water table contours. By definition, this is the direction of groundwater flow. The difference in elevation between two points on this line divided by the distance between them will give you the groundwater gradient.

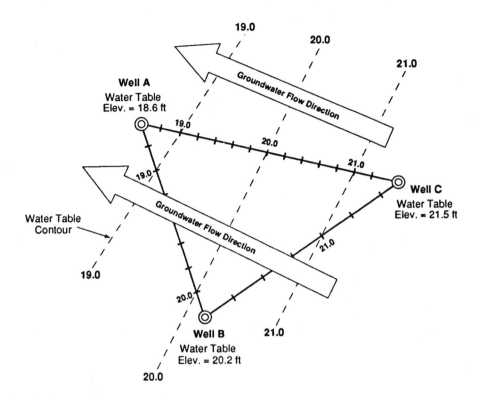

Figure 3-5 Determining Groundwater Flow Direction

Porosity is a concept we have mentioned earlier. It is the percentage of empty space in a soil. In sand and gravel aquifers, porosity generally ranges from 20% to 40%, but can be higher in limestone aquifers.

Now that we have defined all of the variables, we are ready to calculate the movement of groundwater. Let's assume that we are looking at the aquifer illustrated in Figure 3-4. We have already calculated the gradient. Let's assume that the hydraulic conductivity of this aquifer is 150 feet/day and that the porosity is 30% (or 0.3). Using Darcy's equation, the seepage velocity would be:

$$\frac{(150 \text{ ft./day})(0.0025)}{0.3} = 1.25 \text{ feet/day}$$

This is a fairly typical rate of flow of groundwater in a sand and gravel aquifer.

It is a deceptively simple calculation. The only other thing you need to know is that there are a lot of assumptions inherent in this calculation, as there are in all calculations dealing with groundwater. In order to apply Darcy's equation appropriately, you need to know what these assumptions are. The primary assumptions are that the aquifer is simple and uniform and that the aquifer properties that have been chosen are correct. Part of what makes hydrogeology an inexact science is that aquifers are often far from uniform and aquifer properties can be difficult to determine with a lot of certainty. One of the most important things hydrogeologists must come to understand are the limitations of the science they are using. It is a humbling fact of life.

C. WELLS AND GROUNDWATER FLOW

Let us now look at what happens when a water supply well is pumping in an aquifer. Figure 3-6 shows the shape of the water table near a well. Before the well is installed and pumping there is a natural slope, or gradient, to the water table. Once the well starts pumping, this gradient changes near the well.

When water is pumped out of the well, the water pressure (or head) at the well screen is lowered and groundwater begins to flow into the well. The water table in the vicinity of the well lowers at a rate which is determined by the physical properties of the aquifer and predicted by the equations of groundwater flow. This decline in water levels due to pumping is referred to as *drawdown*. If the pumping rate is constant, the rate at which the water table drops will gradually decrease until the water level no longer declines or declines only very slowly. At this point the water levels are said to have stabilized under the pumping condition.

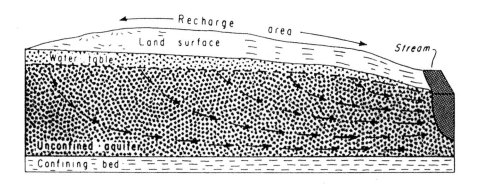

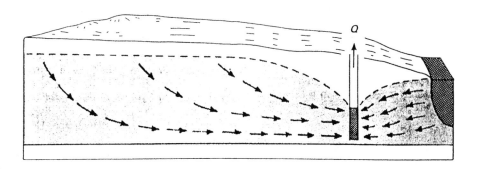

Figure 3-6 The Change in Groundwater Flow Patterns and Shape of
Water Table as a Result of Pumping

The shape of the water table near a pumping well resembles an inverted
cone. In fact, the drawdown is referred to as a *cone of depression*. The rate
of drawdown and the shape of the cone of depression can tell hydrogeologists a lot about the nature of an aquifer. Conversely, if the attributes of an
aquifer are known, the hydrogeologist can predict the drawdown which
results from the pumping of a well.

❹

❋ ❋ ❋ ❋ ❋ ❋

Groundwater as Drinking Water

S
o far we have discussed, in a general way, what groundwater is, how it occurs and how it moves. The next step is to make a more personal connection to groundwater. The study of groundwater would be of only academic interest to geologists, hydrologists and wetland ecologists if it were not for the fact that so many people rely on groundwater every day as the primary source of their drinking water. In this chapter we will be looking at how groundwater gets from the aquifer to the faucet in your home.

Groundwater is supplied to consumers in three primary ways: 1) a domestic water supply well for an individual home, or group of homes, 2) a public water supply system or, 3) public or private springs. Water can be obtained from springs either directly (by filling up your jug at the spring) or by purchasing bottled spring water. We will look at each of these groundwater sources in some detail.

A. DOMESTIC WELLS

If your home is not connected to a public water supply system, you are almost certain to have a domestic well. Approximately 50% of all single family homes in the U.S. obtain water from an on-site well. Domestic wells are more common in rural areas where houses are too few and far between to justify the development of a public water system. Since there are relatively few health safeguards with respect to individual domestic wells, it is important for owners of domestic wells and consumers of domestic well water to have some understanding of their water supply and the potential water quality hazards associated with them.

Let's look first at the basic components of a domestic water supply system. These are illustrated in Figure 4-1.

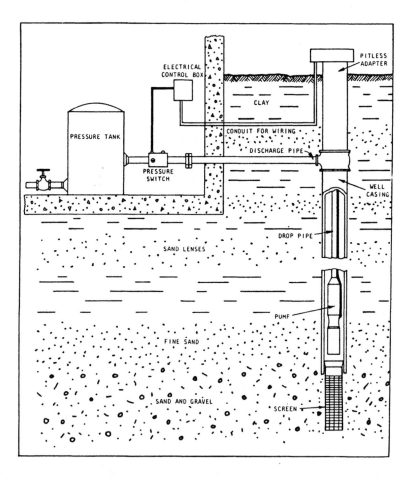

Figure 4-1 Components of a Domestic Water Supply System
(Gibbs, 1973)

The first thing to notice about the well shown in Figure 4-1 is that it is in a sand and gravel aquifer. This means that it will be necessary to have a sturdy *well casing* in order to maintain an open hole. A competent well casing is also important to keep surface bacteria from entering the water. The well casing for a domestic water supply well generally ranges from two to six inches in diameter. In our example in Figure 4-1, a *well screen* is also necessary in order to keep sand and gravel from being sucked into the well. The purpose of a well screen is to allow water to enter the well while keeping the soil particles out.

A domestic well drilled into bedrock does not usually require a well screen because there is not an abundance of soil particles which could be drawn into the well. A bedrock well also does not usually require a well casing within the rock as long as the hole drilled through the rock is stable. A bedrock well will, however, require a well casing for any portion of the hole that is drilled through soil.

The typical domestic well has a submersible pump which is set somewhere below the water table, but not necessarily at the bottom of the well. An electric line and water hose run up the well, through the ground (hopefully below the frost line) and into the house. There is a pressure tank in the house which stores water so that the pump doesn't need to go on every time someone turns on the faucet. The tank is automatically filled from the well once the water pressure in the tank goes below a certain level.

Most domestic wells in either bedrock or sand and gravel aquifers are drilled using one of a variety of drilling techniques such as cable tool, mud rotary, air rotary or reverse rotary. The drilling technique used at a particular site usually depends on the depth of drilling and the type of material to be drilled through. The drilling equipment is typically built onto and hauled around by large trucks, although some drill rigs are fitted onto smaller all terrain vehicles.

Two other options for wells in shallow sand and gravel aquifers are *jetted* or *driven wells*. A *jetted well* is installed by "jetting," or injecting, water into the aquifer. The rapid movement of water in the jetting process removes the soil and makes a hole for the well as it goes in. You can see a similar effect on a smaller scale if you force a garden hose with rapidly running water into the ground. A *driven well* is composed of a small diameter pipe (usually less than 2 inches) with a well screen which is hammered into the soil, either manually or with a powered driver.

Of course, the old-fashioned type of domestic well is the hand *dug well*. Most domestic wells were dug wells up until the early 20th century. Needless to say, these were rather messy affairs. Everything goes pretty smoothly until you reach the water table. Then it becomes very difficult to keep a large diameter hole (at least large enough to work in) open. A dug well is rarely a practical or economic option for a new domestic well. But there are probably plenty of old dug wells still in operation. The primary disadvantages of a dug well are that they are shallow and usually not sealed and therefore vulnerable to contamination from a variety of sources.

B. PUBLIC WATER SUPPLY SYSTEMS

The primary advantage of being connected to a public water supply system is that there is safety in numbers. A homeowner may not have the resources or know-how to properly maintain a water supply well, to frequently test the water quality or to strategize a groundwater protection plan. But a public water supplier can do all these things, and in fact is required to by law. A public water supplier can't afford to be without water due to a power outage or a mechanical malfunction, so equipment is usually well maintained and there are usually backup systems. The water quality of public water supply systems is regularly subjected to a whole battery of water quality analyses. And finally, each public water supplier must take steps to protect the groundwater supply from contamination.

Unlike domestic water systems, the consumers of water from a public water supply seldom know much about the components of that system. They don't have to because someone else is paid to make sure that good quality water comes out of their tap. Although every system is somewhat different, we will discuss some of the common components of a public water supply system. Some public water supply systems have a combination of groundwater and surface water sources. We will be primarily addressing the groundwater sources and how they relate to the water delivery system.

A public water supply system, in its simplest form, consists of one or more water sources and a water delivery system. In addition to these primary components there may also be a water treatment facility and one or more water storage tanks. A schematic diagram of a typical water supply system is shown in Figure 4-2.

THE SOURCE: PUBLIC WATER SUPPLY WELLS

The source may be one or more wells, possibly in combination with a surface water source. A public water supply well is usually substantially larger than a domestic well. A household can get by on a well that supplies only a few gallons per minute, but most public water supplies will need hundreds or thousands of gallons per minute. A public water supply which taps a bedrock aquifer may be simply a larger version of the domestic-type well we described earlier. In sand and gravel aquifers, three types of wells are common: *naturally developed wells*, *gravel packed wells*, and *tubular wellfields*. These well types are illustrated in Figure 4-3.

A naturally developed well consists of a metal casing and well screen, ranging in diameter from several inches to several feet. The well screen is directly in contact with the aquifer. Generally, the width of the well screen openings are chosen so that the smaller soil particles of the aquifer material just outside the screen will be pulled into the well and the larger particles will be left behind. This process of displacing and drawing in the small particles is called "well development." It is done shortly after the well is installed. Water is alternately drawn out of the well at a high rate and then surged back into the well. This helps to loosen the soil particles just outside the screen. Once the well is fully developed (a process that might take hours or days) the aquifer just beyond the well screen will be composed of only the coarser fraction of soil particles. The development process helps improve the efficiency of the well and reduces the likelihoodthat soil particles will be drawn into the well during normal pumping.

A gravel packed well consists of a metal casing and well screen (typically 8 to 24 inches in diameter) which is placed within a larger diameter hole (typically 16 to 48 inches in diameter). The space between the

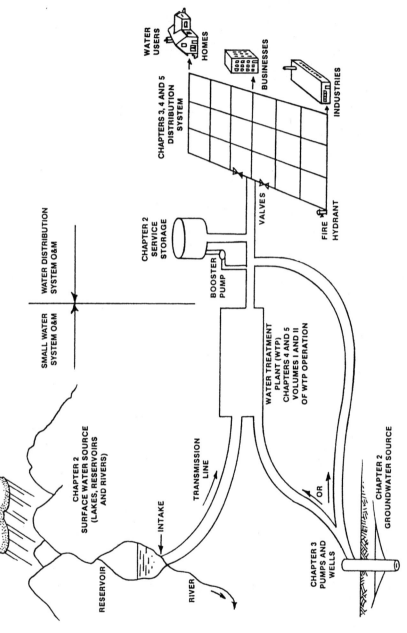

Figure 4-2 Components of a Public Water Supply System
(Copyrighted material from Small Water System Operation and Maintenance.
Reprinted with permission of California State University, Sacramento Foundation)

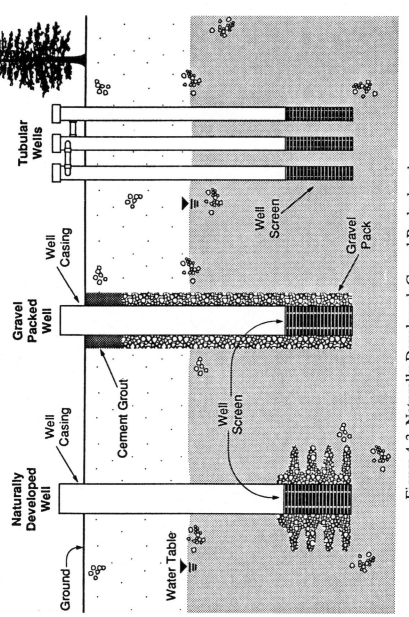

Figure 4-3 Naturally Developed, Gravel Packed and
Tubular Wells

well and the outside hole is then partially backfilled with well-sorted gravel or sand of a very specific grain size. This is the so-called "gravel pack." The gravel or sand "pack" material is like an extension of the screen. It allows water to move into the well freely but maintains a structure around the screen. It is similar to the situation you would hope to have after installing a successful naturally developed well. However, the gravel pack technique eliminates some of the guesswork and usually assures a higher yield. The trade off is that it takes considerably more effort and money to install a gravel packed well.

The term "tubular wellfield" usually refers to a series of small diameter wells (2.5 inch is very common in the northeastern U.S.) which are connected together in a manifold and then pumped by a vacuum pump (see Figure 4-3). Each well individually could only be pumped a few tens of gallons per minute, but if enough of these wells are strung together it may be possible to obtain hundreds of gallons per minute. In thin aquifers (usually less than 30 feet) tubular wellfields can sometimes obtain more water than larger diameter wells. The primary disadvantage of tubular wellfields is that they usually require a lot of maintenance. There are many old tubular wellfields in New England but not many of these wells have been built in the last 20 years. They may, however, make a come back as water suppliers turn to more marginal aquifers in their search for additional water supplies.

WATER SUPPLY TREATMENT

Often, the water which is pumped from a public water supply well is perfectly suitable for drinking straight out of the aquifer. Unlike water obtained from surface water reservoirs, there is usually no need to treat groundwater before it is pumped into the water supply system. There are, however, instances when some kind of treatment is necessary or desirable. If this is the case, the treatment chemicals are generally injected into the water at the *wellhead* itself, as the water is being pumped into the water system. Later on we will be discussing various treatment alternatives for specific types of groundwater contamination. At this point we will simply outline some of the common treatments for well water.

Fluoride is often added to the water in a public water system. This naturally occurring mineral has been found to be effective in reducing tooth decay, especially in children. It is added to the water solely for this purpose. Some studies show that fluoridation of water can reduce the number of cavities in children by 50%. Fluoridation has been hailed as one of the most cost effective public health programs of all times. However, the addition of fluoride to drinking water is not without controversy. Like many chemicals, fluoride can be toxic at higher doses. Detractors claim that there are potential negative long term health impacts even at much lower doses. Nevertheless, fluoride has been added to a large number of water supplies for more than forty years and the majority of health experts agree that carefully controlled fluoridation is safe.

Drinking water obtained from surface water sources is always treated with chlorine or some other disinfection process to eliminate the possibility of bacterial infection. Groundwater, on the other hand, is generally free of harmful bacteria and therefore this type of treatment is usually not necessary. Occasionally, however, if a well is very shallow or is located near a surface water body or other source of bacteria (such as a septic system) the water may be treated with chlorine as a precaution.

In the glacial aquifers of the Midwest and northeast, the groundwater is sometimes quite acidic. This is due partially to the natural acidity of rain (with some manmade extra acidity) and partially to the natural chemistry of the aquifer materials. Groundwater pH (a measure of acidity) of less than 6.0, or even 5.0, is not uncommon (a pH of 7.0 is neutral). This acidic water itself is not particularly bad for you. For instance, most people wouldn't think twice about downing a cola soft drink with a pH of around 2.0. The danger is what may happen to the water as it makes its way through the pipes which lead to your faucet. The acidic water can pick up high concentrations of metals from copper pipes and lead solder in home plumbing (the water mains which actually deliver the water *to the home* do not usually have copper or lead). High concentrations of copper or lead, and numerous other heavy metals, are health hazards. Therefore, if the water obtained from a well is quite acidic, some public water supply systems will raise the pH levels of the water before it enters the system. This is done by adding a basic compound such as sodium hydroxide, potassium hydroxide or lime (calcium hydroxide).

THE DISTRIBUTION SYSTEM

Treated or untreated, groundwater is pumped from the well and into the water distribution system. The water distribution system is a network of pipes of various sizes which convey the water to the houses and businesses in the community. Ordinarily the water mains are buried several feet below each street. Service lines tap into the water main to carry water to each building and house. The location of the water main can often be inferred by the presence of fire hydrants.

Another important part of the water distribution system is the water tank, often referred to as a standpipe or reservoir. A water tank serves the same purpose as a water tank in a home that has its own well. It holds a large quantity of water so that the public water supply well doesn't have to go on and off continuously in order to meet demand. It also stabilizes the pressure in the system and provides a reserve of water for fire protection and peak demand. The well can pump water into the system and the water which does not go immediately to meet demand is pumped to the water tank for storage. Water tanks are generally built up high or placed on a hilltop in order to provide adequate water pressure.

C. SPRINGS

There is a mystique about springs which probably dates back to the days when our earliest ancestors dipped their hairy hands into the clear, cold water which bubbled up from a rock crevice. Tribes and communities were built around springs and depended on them for their existence. Springs are a potent metaphor of the mysterious life-giving power of the earth. They are the backdrops for innumerable myths, fairy tales and legends. Even today there is a widespread belief in the purity and curative powers of spring water.

But the fact is that spring water is simply groundwater that has naturally made its way to the land surface. There is nothing that makes spring water any better than water that you might get from a well. Perhaps spring water is more prized than well water because it is so clearly a gift. You don't have to pump or do anything, the water is just there. But spring water is just as vulnerable to contamination as any groundwater. Actually it is more vulnerable. Since the water is right there at the surface it is subject to bacterial contamination, just like all surface water. In fact, spring water that is used as a public water supply source is regulated by the Environmental Protection Agency and many state environmental agencies in the same way as a surface water reservoir.

What causes spring water to spring from the ground? The answer is fairly simple, really. It is because the water table or piezometric level of an aquifer has reached the ground surface. There are many ways that this can happen. A typical example is shown in Figure 4-4. You might think of it

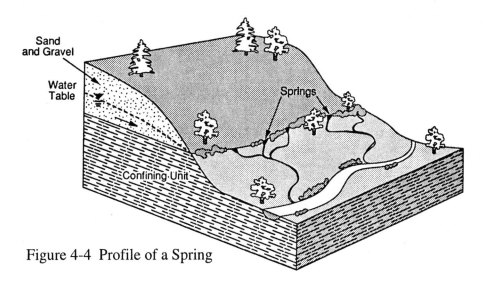

Figure 4-4 Profile of a Spring

this way - water is always looking for the easiest way to flow downhill. In the case of a spring, the easiest way for it to flow downhill is to flow out of the aquifer and onto the ground where it becomes a small stream. In this example, the downward movement of groundwater is limited by the confining unit.

Springs can be a very convenient source of groundwater because it is unnecessary to pump the water out of the ground. However, as mentioned earlier, a spring may be more vulnerable to contamination. Personally, I would not go filling up plastic jugs of water at a spring unless I knew a good bit about where the water was coming from and was sure that it was safe. You certainly can't assume that just because the water is coming from a spring that it must be pure, even if the water is crystal clear, cold and sweet to the taste. You can't see or taste most contaminants.

So if spring water is not necessarily better than any other kind of water, why is bottled spring water such a major industry? Because of the myth that "spring water" means pure, healthful, unadulterated water. The truth of the matter is that there is no guarantee that the bottled spring water that you buy in the store is any better than the water that flows from your tap. There isn't even a definition for "spring water" that all states agree upon. By some definitions any groundwater is considered spring water. According to others, the water must naturally flow out of the ground. The bottom line is that many consumers may be paying a premium for bottled spring water that is no better, or maybe not even as good, as the water from their home tap.

❺

�֟✿✿✿✿✿

Finding Groundwater

The good news is that groundwater can be found just about anywhere there is: 1) at least a moderate amount of rainfall and 2) soil or rock capable of taking in water. This means that groundwater can be found under almost any piece of land on the planet, though it may be scarce in arid climates. However, this does not mean that the groundwater is going to be easy to get at or that there is going to be enough to suit your purpose, or that it is even going to be drinkable.

If you are going to install a well for your home, you will not want to have to get a National Science Foundation grant to pay for the drilling, so a well more than five or six hundred feet deep is usually out of the question. If you bathe regularly and wash your dishes your family will probably need a groundwater source which will provide two or three gallons per minute.

In this chapter we will talk about locating groundwater supplies for drinking water purposes. At first we will address the needs and concerns of the individual homeowner (or prospective homeowner) who is looking for a site for a domestic well. This will be a relatively nontechnical discussion. In most cases, having a Ph.D. in hydrogeology would not help much in the search for an optimum location for a domestic well. Most homeowners have limited land to work with and little money to spend on hydrogeological research. We can, however, offer some insight into the process and some hints to help increase the potential yield of a domestic groundwater source.

Next, we will take a look at how groundwater resources are located for larger public water systems. This is where the science of hydrogeology really plays a major role. Whereas the development of a domestic water supply source might be feasible in almost any geologic formation, a large public water supply source is going to be found only in particular types of geologic formations - those that form major aquifers. We will talk about how these geologic formations are identified. We will also talk about the process of elimination which leads to the siting of a public water supply well.

In many parts of the country, homeowners, farmers and ranchers still turn to dowsers (also known as water witches) to help them locate a site for a water supply well. Dowsing is the ancient and thoroughly unscientific art of locating water by means of a dowsing rod, usually a forked stick. Although scientific research has not supported the claims of dowsers, many people still rely on this folkloric technique to locate underground water sources. We will present a discussion of dowsing from the point of view of groundwater science.

A. FINDING DOMESTIC GROUNDWATER SOURCES

Depending on where you live, finding a domestic groundwater source, a well to serve your home, may be the easiest thing in the world or it may be virtually impossible. For instance, in the eastern, midwestern and northwestern parts of the United States it would be rather unusual if a well were drilled up to 500 feet deep and an amount of water sufficient to supply a home was not found. In most areas of the northeast, sufficient water for home use can be found at depths between 100 and 300 feet. On the other hand, in parts of the arid southwest the chances of drilling a well into potable water might be slim to none. But no matter what part of the country you live in, there are steps you can take to improve your chances of locating a viable domestic water well.

The most important thing to keep in mind when looking for a domestic groundwater source is that groundwater is a local resource. The nature and occurrence of this resource is very much dependant on where you happen to live and what the local geology and climate is. If you were looking for a well site capable of producing a large quantity of water in, say, eastern Massachusetts, a hydrogeologist could give you some advice about the types of rocks and geologic formations to look for. But that same advice would not necessarily be of any use to someone on the coast of Maine or the Ozark Mountains of Arkansas or the panhandle of Florida. That advice might not even help someone find water in other parts of Massachusetts.

In order to increase the chances of locating a good well site, you'll have to do some investigating into your local aquifers. You will need to do a little research. The best place to start is to find out where the good producing wells are in your neighborhood or town. What do these wells have in common? Are they at a similar depth? Are they in a particular type of rock or soil? In a particular locale? Take this opportunity to talk to your neighbors, and possibly the local Board of Health. They sometimes keep records of private wells. If so, it might be worthwhile to spend some time looking at those records. Check the well locations, drilling depths and water quality. If there are patterns in certain areas of town, this might help you learn what to expect.

It also helps to learn a little bit about the geology of your part of the world. You could take this as far as you want. At a minimum, it would be helpful to identify rock types or soil types which are known to produce water in your area. Or perhaps there is a productive regional fracture zone which you can trace on a geologic map. The best resources for geologic information are maps published by the U.S. Geological Survey (USGS). There are many types of maps which the USGS produces which may be helpful including bedrock geological maps, surficial geological maps and river basin studies. If the USGS has investigated the river basin you are located in, the report may include a water resource map (sometimes called a groundwater availability or favorability map). These maps indicate the areas within the basin that are favorable for groundwater development and may even indicate what well yields might be expected. An example of a USGS water resource map is shown in Figure 5-1. This map identifies unconsolidated aquifers with low, moderate and high groundwater yields. Appendix A contains more information on helpful USGS publications and how to order them. Some state agencies also conduct water resource mapping, often in association with the USGS. Included in Appendix A is a state by state listing of agencies which might have helpful groundwater information.

Figure 5-1 An Example of a USGS Groundwater Resource Map

Another potentially good resource is a local water well driller. Find a reliable driller with a couple of decades experience and you have a gold mine of information on your local aquifers. The driller will know where the good producing wells have been drilled and where you might expect to find water quality problems. Well drilling is one profession where experience really counts for a lot, especially if you live in an area where finding a groundwater source is not easy.

On the other hand, you may not have much of a choice as to where your well can go. You may just have a half acre lot. Once you put a house in the middle and measure off the required setback distances from the building and your septic system (check with the local Board of Health or Building Inspector on this), there may not be much room left. Don't worry. Unless you live in an arid part of the country, you still have a good chance of finding groundwater. The advice given above are methods for improving your chances of locating a good well site, but there still is no substitute for drilling a hole and finding out what's down there. As we've said before, hydrogeology is not an exact science. It is difficult to predict the occurrence and movement of groundwater with a high degree of accuracy, especially the occurrence and movement of small quantities of water. There is simply no more reliable method for determining what is occurring below the ground than drilling a hole. So my advice is: do whatever you can to optimize the location of your well, then just drill - and keep your fingers crossed.

B. FINDING PUBLIC WATER SUPPLY SOURCES

As discussed earlier, it is much more efficient to develop, maintain and assure good water quality from a large public water supply well than it is to develop, maintain and assure good water quality from a lot of separate domestic wells. But in order to develop a public water supply well you need to have an aquifer which is capable of producing a relatively large amount of water from one location. Although a family might get by on a well that produces one or two gallons per minute, a public water supply well may not be economically worthwhile if it could not produce more than 100 or 200 gallons per minute. A well producing over a thousand gallons per minute would be considered a large water supply well. It may be a little difficult to imagine how much water this really represents, but just consider that a kitchen faucet probably puts out on the order of two gallons per minute (depending on the water pressure). Your local fire department might get a thousand gallons per minute out of a fire hose.

Unlike the low-yielding aquifers that are often tapped for small domestic water supplies, aquifers capable of producing these larger yields are actually difficult to hide. In order to capture and store large quantities of water, the aquifer will be large enough for a geologist to map as a relatively distinct geologic unit. If the aquifer is shallow, it will probably be identifiable from the ground or from surficial geologic maps. Even if the aquifer is buried deep in the ground it will still be relatively conspicuous to a geologist with the proper data.

In fact, it is probably fair to say that most of the productive and accessible aquifers in the United States have been identified, if not named. Therefore, locating a source for a public water supply well is not exactly like exploring totally unknown territory. It's more like fishing. Fishermen obviously know where the river is. From experience or research they can get to know where the fish are biting. Then it's up to a good lure, patience and luck. In groundwater explorations, hydrogeologists usually know generally where the major aquifers are. Some research can help them narrow down the most promising sites. Then it becomes somewhat of a scientific fishing expedition. And when it comes to groundwater explorations, that means drilling.

Finding the optimum site for a public water supply well is a process of elimination. You begin with an area which is the maximum extent of where you want to look. For a municipality, this might be the town boundaries or the limits of the water distribution system. Within that area, the water supplier wants a water source which meets a certain set of requirements. Some typical requirements might include: 1) the site should be capable of producing a certain amount of water on a long-term basis, 2) the site should be free of contamination and remote from potential contaminant sources, 3) there should be enough land available to meet wellhead protection restrictions (which vary from state to state but are generally between 200 and 400 feet from a public water supply well), 4) the site should be accessible for testing and, 5) the site should be either publicly owned or obtainable by purchase or by "eminent domain" (the legal process by which a municipality can force the sale of a property).

As the municipality or private water supplier evaluates the aquifer in respect to each of these criteria, they will begin to eliminate areas of the aquifer which do not meet these criteria and narrow down the potential water supply sites within an aquifer. In the course of this process usually a few sites emerge as having the highest potential.

In order to illustrate this process, let's take the example of a town water department that is searching for a new well site in a sand and gravel aquifer. We'll call this town Springfield. Springfield is a growing town that needs another million gallons per day of water in order to meet the town's water demand for the next ten years. Let's take a look at how this process of elimination we've been talking about might work.

Figure 5-2 is a map showing the extent of the town, some major roads and the extent of a major sand and gravel aquifer that has been mapped by the U.S. Geological Survey. The area they will be considering for a new well source is in that portion of the aquifer which is within the town boundaries. It may have been legally possible for them to develop a well site in an adjoining town but this can get very complicated; legally, politically and administratively.

The state in which Springfield is located has regulations requiring that all public water supplies must be situated on land owned by the water supplier and that the well must be at least 400 feet from all property boundaries. This is the setback regulation which we mentioned earlier. It is intended to provide a protective perimeter around the well. This type of land-use protection in the immediate vicinity of a well is important because any contaminants which are spilled onto the ground could make their way to the well relatively quickly.

What this restriction usually means in practice is that the Water Department must locate a well site on an undeveloped piece of land that is at least 400 feet from any road or structure. Our fictitious town of Springfield is fairly well-developed so this restriction eliminates a relatively large portion of town.

In addition to the political, hydrogeological and regulatory restrictions, the Water Department has developed its own set of criteria for the siting of this well: 1) they want the well to be located within the area which is already served by the water system, 2) for water quality reasons, they do not want the well located in a portion of the town zoned for industrial use, 3) they wisely do not want the new well to be located downgradient of the

Springfield, U.S.A.

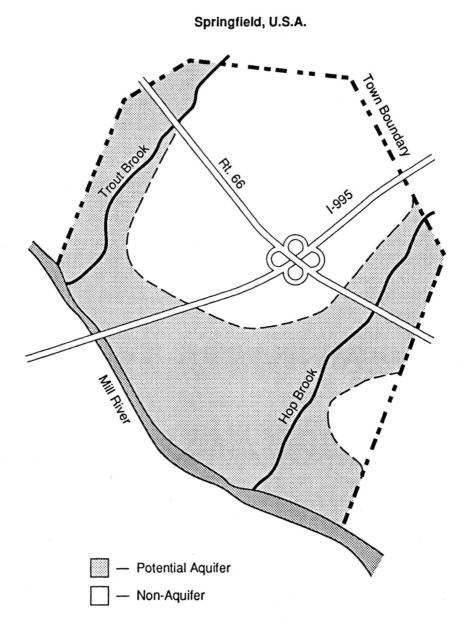

- Potential Aquifer
- Non-Aquifer

Figure 5-2 Potential Aquifer Areas in Springfield, USA.

Springfield, U.S.A.

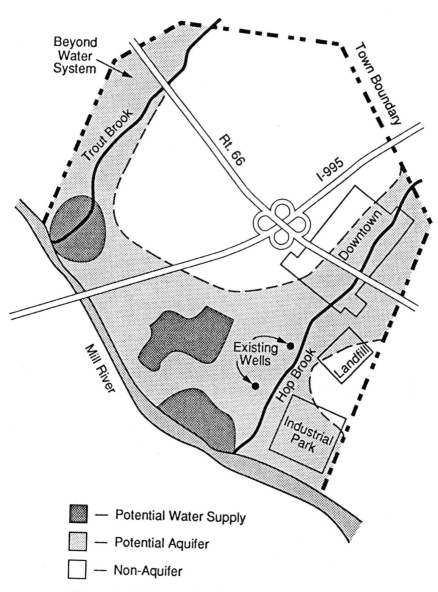

- ▨ — Potential Water Supply
- ▧ — Potential Aquifer
- □ — Non-Aquifer

Figure 5-3 Potential Public Water Supply Sites in
Springfield, USA

town landfill or any other known source of contamination and, 4) they do not want the new well to be within 800 feet of an existing public water supply well. By including these criteria, the Water Department has limited itself to the areas shown in Figure 5-3.

The next step for the Springfield Water Department is to conduct a test well investigation in these chosen areas. Under some circumstances, the public water supplier might just acquire the land and build the final well without any testing at all. But this would only be done if the necessary amount of water could be easily obtained and the water quality was known to be acceptable. It is much more prudent to conduct a test well program. The purpose of a test well program is to measure the potential yield of the site and to sample the water quality before building the final well and pumping station. If the site turns out to be unacceptable or marginal, the Water Department has risked less of an investment and another site could be investigated.

In the northeastern United States, a test well program is most often conducted by drilling 2.5 inch diameter wells. A test well drilling program will usually be conducted on a number of potential sites until one or more favorable sites are located. Once a test well is drilled to the desired depth, it is pumped to determine the potential yield. Often water levels are measured in an adjacent well during the pumping in order to better determine the potential yield of the aquifer. If the yield is adequate, a water sample will be obtained for thorough water quality analysis. If the yield and water quality is acceptable, the site is considered favorable for a public water supply well.

Before the public water supplier builds the final well and pumping station, there is often a further level of testing known as a prolonged *pumping test*. The purpose of a prolonged pumping test is to subject the aquifer to a level of pumping similar to that which will occur when a final water supply well is installed at the site. This means that a well (either the final well or another test well) is pumped at a high rate of flow for several days. This makes it possible to more accurately determine how the aquifer will respond to the pumping of the final well.

A pumping test is especially important when evaluating sites with marginal potential. A small aquifer composed of coarse sands and gravel may look good when pumping small diameter wells but may prove to be disappointing when a larger volume of water is pumped. Also, a higher rate of pumping will draw water from a larger area, and as a result, the water quality may change. In addition, a properly designed and executed pumping test can provide a great deal of information about the properties of the aquifer. This information can, in turn, provide a hydrogeologist with the data necessary to predict regional groundwater flow and the movement of potential contaminants. These types of analysis are very important when trying to calculate a *zone of contribution*, that is, the area which supplies water to a particular well.

A prolonged pumping test is usually conducted by pumping one well at a constant rate for a period of days while measuring water levels in a series of nearby *observation wells*. The observation wells are placed at optimal distances from the pumping well (on the order of 50, 100, 200 and 300 feet, depending on the type of aquifer and the pumping rate). While the test well is being pumped, the water levels in the observation wells will gradually draw down. When the pumping ceases, the water levels begin to recover to their original levels. By plotting the rates of drawdown and recovery, hydrogeologists can calculate aquifer parameters such as *transmissivity* and *specific yield* (or *storage coefficient*). Transmissivity is a measure of how quickly water could potentially move through an aquifer. Specific yield is a measure of how much water is available in a given volume of aquifer.

If a prolonged pumping test is conducted with satisfactory results, both in terms of yield and water quality, the chosen well site will almost certainly be capable of providing a long-term supply of good quality water. The search is over

C. Is Dowsing an Option?

Some people, particularly in rural areas, turn to dowsers (also known as water witches) to help them find water. In many ways this is understandable, since most people have probably heard more about dowsing than they have about hydrogeology. You may even know someone who could tell a story about how a dowser helped someone that they know find water.

Since dowsing is such an integral part of the folklore about groundwater and because many people today still rely on the art of dowsing to find groundwater, I thought it would be important to say something about the subject in this book.

Part of the difficulty of even discussing the subject is the fact that hydrogeology and dowsing operate in completely different worlds and their proponents speak different languages. Hydrogeology operates within the realm of science - the realm of the known, the visible, the quantifiable (although it sometimes seems that there is little about an aquifer which is known, visible, or easily quantifiable). The discipline of dowsing, on the other hand, operates within the realm of folklore and magic, a realm which, to many people, groundwater seems to be more at home in.

It is unusual to find a dowser versed in geology and even more unusual to find a geologist who dowses. This, however, does not prevent adherents of either discipline from denouncing the other at great length. Hydrogeologists and engineers, for the most part, consider dowsing to be a complete waste of time. Dowsers are more divided on their views on the science of hydrogeology. Some dowsers regard hydrogeologists and engineers as colleagues. Others consider them to be misguided and unobservant.

Dowsing is an ancient form of divination for water or minerals (or, more recently, just about any object) usually by means of some type of device. Typically this device is a forked stick often referred to as a divining rod. However, there are many types of dowsing instruments including "L" shaped rods and pendulums. Typically, the dowser holds the divining rod or other device in his or her hand and walks across a property. At the point where the dowser crosses a "vein" of water, the device moves - the divining rod turns downward, the "L" rods cross, or the pendulum makes a certain motion.

The origins of dowsing are uncertain. It may be related to several other forms of divination. One of the earliest descriptions of dowsing comes from mining texts from the Middle Ages. Johannes Agricola, in his text De Re Metallica (1556) describes techniques for locating water and minerals by means of a divining rod. Even at that time the technique was considered controversial and its usefulness was by no means certain.

Many scientific, and not so scientific, investigations have been made into dowsing over the years to try and determine if and how it works. The methods and results of many of these investigations have been rather questionable, especially some of the earlier investigations. However, most of the unbiased and objective studies do not support the claims that dowsing is an effective way to locate water. Many dowsers respond to these studies by stating that scientific studies like these are, by nature, biased against dowsing and prefer to base their claims on a preponderance of anecdotal evidence.

One thing which studies have shown, and which dowsers generally agree to, is that the divining rod itself does not seem to have any properties which are affected by water. It is not an interaction between water and the divining rod which makes the stick move downward. The divining rod is not a water locating machine. The ability to locate water is considered to be a kind of psychic ability on the part of the dowser. This explains why the adherents of dowsing claim they can use the technique not only to locate water but also to locate lost articles and diagnose illnesses. In fact, at the annual conventions of the American Society of Dowsers in Danbury, Vermont, relatively little interest is generated in dowsing for water. There seems to be much more interest in dowsing auras and other occult pursuits.

Water dowsers have developed theories of how water moves through the subsurface which are very different from the hydrogeological concepts presented in this book. Dowsers talk about underground water flowing in "veins." This seems to be a vestige of the application of dowsing to mining. Very often metals occur in veins of rock. The vein concept also corresponds to the familiar movement of water in surface streams. It is easy to see how such a concept could develop, but decades of scientific observation and study have shown that groundwater does not typically flow in underground rivers or veins at all (there are rare exceptions in limestone and some hard rock aquifers).

The dowsers do have one very convincing argument which seems to explain why they still remain popular in this day and age - dowsing seems to work! It is not uncommon for a dowser to claim to have an 80 or 90% success rate in locating well sites. How do the scientists explain that?

Part of the answer is in the first paragraph of this chapter. Small quantities of groundwater are often not all that difficult to find. You can drill a well almost anywhere in the eastern, midwestern or northwestern areas of this country and find enough groundwater for a domestic well. So an 80 or 90% success rate in those areas is not actually all that impressive.

A farmer once proudly pointed out to me how a dowser had located a vein of water within 100 feet of his house. The well pumped 30 gallons per minute easily, just as the dowser had predicted. And there we stood above a deposit of sand and gravel which stretched almost as far as the eye could see. The farmer was a bit skeptical when I told him that he could have drilled anywhere he wanted to on his property and gotten hundreds of gallons per minute. To many people dowsing does not seem nearly as occult as the science of hydrogeology with its obscure mathematical formulations and unseen aquifers created thousands or millions of years ago by phantom rivers and glaciers.

In summary: 1) the dowser's concept of how groundwater occurs and moves is usually contrary to observable phenomena, 2) the large body of evidence from scientific investigations into dowsing fails to substantiate the claims of dowsers and, 3) if dowsing works at all, it is a purely psychic phenomena. If you are going to hire a dowser to find a water supply you might as well have him check your aura, diagnose your heartburn, free your Chevy of evil spirits and choose a stock portfolio for you.

If you are interested in learning more about dowsing, there is a list of references in Appendix A.

6

✽✽✽✽✽✽

The Importance of Water Quality Testing

Most people probably don't know, or care very much, where the groundwater that they drink comes from. But they do generally care about the quality of their drinking water. In spite of this, most people are as unaware of the quality of the water they drink as they are of where it comes from.

Fortunately, the majority of groundwater consumed in the United States is of good quality. According to the U.S. Environmental Protection Agency's (USEPA) 1992 National Water Quality Inventory, the quality of U.S. groundwater is described as "good to excellent." Even so, according to one review of the Federal Reporting Data System (a USEPA data base), almost 10 million Americans drank water from public water systems which violated established drinking water standards (Ainsworth, 1995). The EPA's National Survey of Pesticides in Drinking Water Wells, Phase II Report, estimated that the drinking water of more than 4.5 million people contains levels of nitrates above drinking water limits (USEPA, 1992). The problem areas are primarily in industrial and agricultural regions. But how would an individual consumer know whether their drinking water fell within the good to excellent category or was tainted in some way?

As you can gather from our earlier discussion of the hydrologic cycle, water is constantly moving through our environment, every day, no matter where we happen to be. As water moves through the hydrologic cycle it mixes with and dissolves, to some degree, whatever comes in its path. Water is referred to as the universal solvent. In the air, water picks up gases and chemicals, such as the nitric and sulfuric acids found in smog. Most rain is naturally acidic anyway, but air pollutants tend to make the rain even more acidic. Rain pours down on the earth and very, very gradually erodes and dissolves everything that is solid - from mountains to buildings. This is a slow process that has been going on for untold eons and not something that you are going to notice on a day to day basis.

Groundwater that has been in the ground for a long time becomes naturally enriched in minerals that the water has dissolved from the rocks and soils. These dissolved minerals flow in solution in the groundwater and surface water to the sea. In the sea, the solution becomes more concentrated as water evaporates and leaves the minerals behind, just as happens in salt lakes. This is why ocean water is so "salty" or enriched with minerals.

This also is a natural process which has been occurring for billions of years. The net result is that the quality of the groundwater in a given area is conditioned by the chemical makeup of the rocks and soil through which it flows. Groundwater in a limestone terrain will likely have a very high pH (that is, it will be *basic* as opposed to *acidic*) and the water will be considered *hard*. We will discuss these terms in more detail later. Groundwater in granite terrains, on the other hand, will probably be acidic, and may contain natural contaminants associated with certain types of granite, such as arsenic or radon.

Add to this mixture the hundreds of thousands of manmade chemicals which are produced every day. Many of these chemicals do not simply disappear once they have been "used." If they are released into the air, discharged to surface water bodies or dumped on the ground, they may become a part of the hydrologic cycle and could potentially contaminate groundwater.

The big question is: what are the potential health impacts of consuming these chemicals and natural minerals? Each of the hundreds of thousands of chemicals and minerals has a potential health impact associated with it. That impact may be harmful or healthful, dramatic or undetectable. Many of the minerals commonly found in groundwater are not only healthy, but are essential nutrients. Other naturally occurring and manmade substances are toxic above a certain concentration. Some represent a measure of risk at any concentration. Some are healthful within a range of concentrations but are toxic at higher levels. For example, earlier we mentioned fluoride, a mineral which is naturally present in nearly all groundwater. At low concentrations in water, fluoride is an effective cavity fighter. At higher concentrations, fluoride can cause dental fluorosis, a condition where teeth turn yellow, brown or black and become brittle.

Suffice it to say that the issue of water quality is vastly more complex than most water drinkers can imagine. The conflicting reports and data can be used to justify the positions of those who think that all tap water is poison, as well as those who think that environmental protection is a government boondoggle. Most informed people choose positions somewhere between these extremes.

In the next few chapters we will try to present a balanced view of water quality. The tendency of journalists and public officials in the past has been to oversimplify the issue. This has led to some misunderstandings and mistrust. The fundamental issue to be grappled with here is one of risk. There are risks associated with every act: driving a car, walking down the street, investing money, going on a date. You can learn, more or less, the risks associated with these actions. But the chances are that most of us know very little about the risks inherent in our drinking water.

The emphasis of these next chapters will be on the quality of groundwater sources. Many of these water quality issues will apply to all sources of drinking water, including surface waters. However, water quality issues which apply primarily to surface water sources will not be discussed in any detail. If the water utility which supplies your home has a combination of groundwater and surface water sources, you may need to be concerned with some water quality issues which are not discussed here.

A. GETTING THE FACTS ABOUT WATER QUALITY IN PUBLIC WATER SYSTEMS

If your water is supplied to you by a water utility, by federal law that water has to be tested regularly for a wide variety of potentially harmful contaminants. If the concentrations of these water quality parameters exceed a certain level that is set by the EPA, known as the *Maximum Contaminant Level* (MCL), the water purveyor will need to address this potential problem. This provides a certain level of security for those of us on a public water supply. What water quality parameters must be tested for, how often these tests must be made and what levels of contaminants are considered safe in a public water supply system has been set nationwide by the EPA in accordance with the provisions of the Safe Drinking Water Act passed by Congress in 1974 and amended in 1986.

Let's consider for a moment the concept of the MCL. If the concentration of a given contaminant is at or above the MCL value, the EPA would consider the water unsafe. This does not necessarily mean that the water utility would be forced to immediately shut down the water supply. The immediate shut down of a water supply would only be done if the level of contaminant were high enough to pose an immediate health hazard. Often, the presence of a contaminant in drinking water at a level above the MCL does not pose an *immediate* health threat. The water can usually still be used as a drinking water source, but further testing for the contaminant would be required. If the level of contamination did not naturally fall below the MCL by a certain period of time, then the water would have to be treated until the concentration once again fell below the MCL. In some cases the water source might be temporarily, or permanently shut down.

One of the problems inherent in the concept of the MCL is the fact that it is so black and white. For example, the MCL for the chemical benzene (which is found in gasoline) is 5.0 parts per billion (ppb, which means that for every volume of benzene there are a billion volumes of water). Would it be safe to drink water with a benzene concentration of 4.9 ppb? According to EPA regulations there would be no restrictions on a water supply with this level of benzene. What if the concentration increases to 5.1

ppb? According to the EPA, this would be an unsafe level for a drinking water source. Nothing really magical has happened to this water as the concentration rose 0.2 ppb. The health impacts from these two concentrations are essentially the same. But in order to devise a workable system to safeguard the quality of water supply systems the EPA had to decide what is an acceptable health risk and then choose a cutoff point concentration so that this health risk is not exceeded.

This is easiest to illustrate in the case of substances that are carcinogenic. It is known that the consumption of a carcinogenic substance poses a certain level of risk for contracting cancer. The higher the rate of consumption, the greater the risk of cancer. In many cases there is no lower limit of concentration which will have absolutely no cancer risk. Completely eliminating all of these substances from our environment would be an extraordinarily difficult, if not impossible, task. Since it may be impractical to completely eliminate a given chemical from the environment, the EPA may use a criterion of one additional cancer death in a million as an unacceptable limit. In other words, if one million people drink this water with 5 ppb of benzene in it every day for the next 20 or 30 years, statistically, one of them might die of cancer as a result of that exposure.

This may seem rather cold and calculating. And some could argue that they would prefer drinking water which has absolutely no risk of causing cancer. But the fact of the matter is that most of us would not think twice about engaging in activities which have degrees of risk which are many times higher than that - activities such as driving a car, going to the beach or just living in a city. According to one study (Hird, 1994), each of the following activities increases your chances of dying in any one year by a factor of one in a million: 1) driving 150 miles in a car, 2) living in New York City or Boston for two days, 3) living with a cigarette smoker for two months, 4) riding a bicycle for ten miles or, 5) drinking half a bottle of wine. The point is that each of us is willing to accept some risks, but perhaps not other risks.

In general, the MCL levels set by the EPA are very conservative and even if a given chemical is found in drinking water at a level which is just slightly above the MCL, this might represent only an extremely minor health risk. On the other hand, these minor health risks can be cumulative.

If your drinking water contains fifty chemicals at concentrations which are only slightly below the MCLs (and therefore acceptable by EPA standards), the combined health risks might add up to something quite significant. Another point to keep in mind is that for many chemicals, the health risks are not well understood. Several times over the past 20 years the EPA has revealed that a level which was once considered safe for a chemical is now known to be hazardous because of new results from more detailed or long-term studies.

Most people seem to sleep very well at night without worrying about the quality of their drinking water. By federal law, the water quality of public water systems must be tested and water quality problems which exceed MCLs must be addressed. If you don't own your own well and have confidence in your water utility's ability to address these issues, then you can probably skip the next two chapters. If you are interested in learning more about the quality of your drinking water then read on.

All public water supply systems are required to conduct numerous water quality analyses and submit the results to the proper state authority. These water quality records are public documents and are available for review by any concerned citizen. Very few people actually exercise this right. If you decide to check these records for yourself, keep in mind that your local water purveyor or state agency probably does not have a full time staff dedicated to responding to water quality inquiries. If you are curious and persistent and the local water utility or state agency supplies you copies of the appropriate water quality analyses, you will have to learn how to decipher these documents. For assistance in this, refer to Chapter 9, section C.

B. EVALUATING THE WATER QUALITY OF YOUR PRIVATE WELL

The consumers of publicly-supplied water are far better protected than an individual domestic well owner. Even if those consumers never bother to check the water quality records for themselves, they can be reasonably sure that the water utility is closely monitoring the water quality, as required by law. Most homeowners, on the other hand, are not going to be able to afford the level of testing which a public water supply undergoes, at a cost of thousands of dollars per year. A large percentage of domestic well owners have their water tested only for bacteria and a small number of relatively benign chemicals at the time that their well is installed, and then don't bother to have water tested ever again.

This practice leaves the water supply of millions of people at risk, a fact which may rank as one of the least publicized health threats in America. It is relatively easy to understand how this situation has come about. Most people would think that if water is clear, colorless and odorless, it is probably of good quality. The truth is that most of the toxic and carcinogenic compounds which might reach a water supply well would not be detected in this way. In addition, there are usually no regulations requiring home builders or developers to extensively certify the quality of water for the homes they sell.

If you drink water from your own well and are wondering whether or not you should have your water tested, take this simple test.

1) Look around the area you live in. Are there any of the following within a mile radius of your house: any industry (factory, power plant, high tech firms, metal plating etc.), gasoline stations or other chemical storage, a landfill (in use or abandoned), a dry cleaner or a service garage?

2) Do you live in an agricultural area?

3) Could your well possibly be drilled into granite rock?

4) Do you or your neighbors have a septic system?

5) Is it possible that there are lead pipes or solder in your plumbing system?

6) Does your water sometimes taste or smell "funny" or offensive?

7) Are you someone who does not like to take chances?

If you answered "yes" to any of these questions, then you may want to have your tap water tested. In the next chapter we will be talking about the different water quality parameters you could test for. If you want your drinking water to be as well-protected as a public water supply, it might cost you a thousand dollars per year. This is not a feasible option for most people. A better approach would be to evaluate the specific water quality threats which could possibly impact your well and test for the chemicals associated with those threats. In Chapter 8 we will describe the types of contaminant sources which could threaten the quality of water in a drinking water well. In Chapter 9 we will present a checklist which may help domestic well owners decide on the minimum water testing that they should do.

Just a note of warning before we delve into the issue of drinking water quality in more detail. In the next few chapters you will be reading about many different potential sources of contamination: toxic wastes, spills of cancer causing chemicals, releases of industrial chemicals, pesticides, natural hazards and probably a lot of sources you never even thought about. It is perfectly natural for the simple awareness of these threats to make some people a little uneasy. It's like studying the details of all the possible diseases a body could be exposed to. You might begin to imagine that you have some of the symptoms of those diseases. Be aware of this effect and don't worry too much. The chances are very good that your water quality is perfectly fine. There are certainly enough dangers out there to make any domestic well owner want to test their water quality. But most people who test their water will be reassured by the results. One out of a hundred readers might discover something in their water to be a little concerned about. One out of a thousand might find something alarming. The author would be gratified to know that the information presented in this book prevented a few households from continuing to drink tainted well water.

The Chemical and Biological Players in Groundwater Quality

Contrary to popular belief, there is no single set of standards to define what is good drinking water quality. You cannot simply take a sample of your water, bring it to a lab, have them run it through a machine and give you a slip of paper telling you that your water is good or that it is not good. Life should be so simple. Instead, if you take a sample to a water testing lab you will have to decide among the hundreds of possible chemical analyses which they offer. And when you get the results, you will have to decide for yourself whether or not the water quality is as good as you want it to be.

Although there is no single test which can tell if the quality of your water is acceptable, fortunately you will not have to test for all of the thousands of chemical compounds known to man. There are several reasons for this. Many chemicals are either not soluble in water under normal conditions or they are extremely rare. The EPA and others have narrowed the field of likely drinking water contaminants to a few hundred chemicals and several biological agents. To make this list more comprehensible we have divided these water quality parameters into six categories: 1) microorganisms, 2) common inorganic constituents, 3) heavy metals, 4) synthetic organic chemicals and, 5) radioactive substances.

Public water supplies are tested for virtually all of the water quality parameters discussed in this chapter. If you have your own domestic well and would like your water to be as safe as a public water supply, then you should analyze your water for all of these parameters. If this option is too much for you to swallow (pardon the pun), then refer to Chapter 9 for ways to tailor your water quality testing to include the parameters which are more likely to be a concern at your well.

A. MICROORGANISMS

Although microorganisms, such as bacteria, viruses and parasites, are common water quality concerns for surface water supplies, they are rarely a major concern for groundwater sources. Studies have shown that these microorganisms generally do not survive very long, or travel very far, in the comparatively sterile environment of an aquifer. Therefore, microorganisms do not usually pose a threat to water supply wells *unless*:

1) There is a potential source of these pathogens (such as a cesspool, septic tank or leaking sewer line) relatively close to the well. "Relatively close" is difficult to define because it depends on the nature of the aquifer and the construction of the well. As a rule of thumb, if there is a septic system within 100 feet of your well, especially if you have a shallow well, testing your water for the presence of bacteria would not be a bad idea.

2) Water is "short-circuiting" the aquifer and going unhindered from the ground surface to the well, thereby bypassing the natural filtering process of the aquifer. The most common cause of "short-circuiting" is a crack or leak in the well casing itself, which allows surface water to enter the well directly. In addition, if your well is very close to a pond or stream, surface water may be reaching your well after traveling only a short distance through a gravel lens or bedrock fracture.

Even though contamination of groundwater by microorganisms is relatively uncommon, testing well water for bacteria is a common practice. Most often, if the results come back positive (assuming that the water sample was not contaminated during the sampling process) it is because the integrity of the well has somehow been compromised. If this is the case, the well should be repaired and disinfected. If the problem persists, then it is likely that the contamination is in the groundwater itself, not just the well. The problem may be a nearby septic system. It would be advisable to contact your local Board of Health and get professional assistance.

There are many, many kinds of bacteria which we come into contact with every day. However, water testing laboratories generally provide analyses for two specific kinds: total coliform bacteria and fecal coliform bacteria. Coliform bacteria are usually harmless but if they are found in your water it indicates that harmful bacteria may be present as well. Fecal coliform indicates contamination from the fecal matter of warm-blooded animals. These bacteria can cause nausea, diarrhea and other gastrointestinal distress.

If there are bacteria in your water and it is coming from a septic system, then there may be viruses in the water too. Viruses are much more difficult to test for. On the other hand, it shouldn't even be necessary to test for viruses. If you have bacteria in your water, you should assume that there may be viruses as well and boil it or just not drink it until the problem has been solved.

Protozoan parasites, such as giardia and cryptosporidium, are even less likely to be found in groundwater than bacteria because these organisms are larger and are filtered out by the aquifer materials. These organisms are, however, a major concern for surface water supplies such as reservoirs. There is also a remote chance that giardia or cryptosporidium could enter a well if the well is very near a pond or stream.

B. COMMON INORGANIC CONSTITUENTS

This category of water quality parameters primarily consist of naturally-occurring dissolved chemicals, or *ions*, in groundwater, which are derived from minerals in rocks. Some of these are essential nutrients, like calcium and iron. At the appropriate levels, these minerals are not only harmless, many of them are beneficial. Pure H_2O would not provide you with some elements that your body needs and probably would not taste very good to you. On the other hand, many people consider mineral waters to be healthful. Mineral waters are enriched with the common inorganic constituents we will be discussing here.

These same chemicals can be unnaturally enriched in groundwater by human intervention, to the point when they are considered to be contaminants. It is not always easy to tell if an element or chemical is present in groundwater "naturally" or if it is due to manmade pollution. Each aquifer has a natural background level of these parameters. The only way to determine what level of a given parameter is natural is to test for that parameter over an extended period of time. It is only when background levels are significantly exceeded that it is clear that contamination has occurred.

Table 7-1 lists what we will refer to as the "common inorganic chemicals" along with: their chemical abbreviations, MCLs, Secondary Standards (recommended limits based on aesthetic considerations, not health), possible sources and possible health effects. Most of these parameters should be familiar to you. Note that only a few of these actually have MCLs. This means that there is no regulated upper limit for the concentrations of these water quality parameters.

The MCL concentrations are given in units of mg/l (milligrams per liter). This is essentially equivalent to the commonly used term "parts per million" (ppm). There are a million milligrams in a liter. Sometimes concentrations are given as micrograms per liter (ug/l). Since there are a billion micrograms in a liter, this unit is essentially equivalent to parts per billion (ppb).

Included in this group are two water quality parameters which do not fit easily into any category of chemicals, but which are usually associated with the common inorganic minerals and chemicals. These water quality parameters are *pH* and *specific conductivity*. The pH of a liquid is a measure of how acidic or basic that liquid is. A pH value less than 7.0 pH units means that the liquid is acidic, above this value the liquid is considered basic. In and of itself, pH is not an indicator of water quality since some waters are naturally acidic and some are naturally basic. The pH value can, however, provide an indication of whether or not the water is corrosive and therefore could more easily dissolve metals, including your water pipes. Water with a low pH is more likely to be corrosive and enriched in dissolved metals such as lead and copper.

Specific conductivity is a measure of the ability of the water to pass an electric current. Waters which are enriched in dissolved ions have a higher specific conductivity because these ions aid in the conductance of electricity. It is a very simple test to perform and can provide an overall assessment of the concentration of dissolved chemicals in the water. The results are typically given in microsiemens (μs). Mineral waters might have a specific conductivity of several hundred μs (sea water is approximately 35,000 us), whereas a purified, de-ionized water would have a specific conductivity of almost 0 us. *Total Dissolved Solids* (TDS) is another general measure of dissolved constituents in water. As a general rule of thumb, the specific conductivity multiplied by 0.65 gives a rough approximation of the level of total dissolved solids in parts per million. This is only a rough approximation because some ions conduct electricity better than others.

Each of the remainder of the chemical parameters in Table 7-1 represents chemical *ions*. Ions are electrically charged atoms or groups of atoms. An ion with a positive charge is known as a *cation* and an ion with a negative charge is an *anion*. We are getting dangerously close to a chemistry lesson here. Let's just say that within rocks and soils the positive and negative ions are bound together to form minerals. When water dissolves a mineral it loosens those bonds and the ions attach themselves to water molecules. The mineral is then said to be "in solution."

Let's take a simple example. What we think of as salt is composed primarily of sodium chloride, NaCl. When salt is dissolved in water the solid seems to disappear. This is because the Na^+ and $Cl-$ ions (cation and anion, respectively), which were previously bonded to each other, have now bonded with the water molecules. When a lab tests for sodium chloride in your water, they test for Na^+ and $Cl-$ separately, in solution.

Table 7-1 Common Inorganic Water Quality Parameters

PARAMETER	SYMBOL	MCL (mg/l)	SECONDARY STANDARDS	POSSIBLE SOURCES	POSSIBLE EFFECTS
Alkalinity	--	NA	NA	N.O.,landfills	Possibly same as hardness.
Calcium	Ca	NA	NA	N.O., landfills	
Chloride	Cl	NA	250	N.O., landfills, road salt	Salty taste, corrosion of pipes
Hardness	--	NA	NA	N.O., landfills	Reduces effectiveness of soaps.
Iron	Fe	NA	0.3	N.O., landfills	Metallic taste, laundry staining
Magnesium	Mg	NA	NA	N.O., landfills	
Manganese	Mn	NA	0.05	N.O., landfills	Taste, laundry staining
Nitrate (as Nitrogen)	NO_3	10	NA	landfills, septic systems, farms	Methemo-globinemia
pH	pH	NA	6.5 - 8.5	N.O., landfills	Low pH: Metallic taste, corrosion High pH: slippery feel, soda taste
Potassium	K	NA	NA	N.O., landfills fertilizers	
Sodium	Na	NA	NA	N.O., landfills road salt	Possibly increased blood pressure
Specific Conductance	--	NA	NA	N.O., almost any source	
Sulfate	SO_4	500	250	N.O., landfills	Salty taste, laxative effects
Total Dissolved Solids	TDS	NA	NA	500	N.O., almost any source	

N.O. = Natural Origin

The most abundant dissolved inorganic constituents found in ground-waters are calcium, magnesium, sodium, potassium, bicarbonate, sulfates, and chlorides. Minor constituents of groundwater include iron, carbonate, nitrate, fluoride and boron.

If you live in an area with carbonate rocks, such as limestone, the chances are good that your water will be "hard." Hardness is a rather un-scientific measure of water quality. It is the measure of how easily water can produce bubbles when soap is added. The harder the water, the more difficult it is to make bubbles, and the more difficult it is to clean anything. Hardness is primarily caused by the presence of calcium and magnesium, therefore a laboratory test for hardness often consists of testing for the combined concentration of calcium and magnesium. Carbonate rocks can be rich in both calcium and magnesium. There are no EPA health-based drinking water limits associated with hardness.

In granitic terrains, the groundwaters tend to be more acidic. In these acidic groundwaters iron and manganese tend to be more common than calcium and magnesium. These constituents also are not considered to have health risks associated with them. They can, however, be a nuisance if the concentrations are sufficiently high. Groundwaters with high concentrations of iron and manganese can cause staining of clothes and sinks. This is because when the waters become exposed to the air they become oxidized. Iron and manganese will precipitate when oxidized; that is, they will come out of solution. Everyone has seen oxidized iron. It is rust. Precipitated iron may cause red or brown staining. Precipitated manganese will stain black or gray.

Nitrates are usually only a minor constituent of groundwater, but they are a common contaminant. Nitrates are one of the few common inorganic constituents which have an MCL. High nitrate levels in groundwater are usually an indication of contamination by animal (including human) waste or agricultural fertilizers. Nitrates are a matter of concern because levels above approximately 10 mg/l nitrate as nitrogen can cause a relatively rare blood disorder in infants called methemoglobinemia, the so-called "blue-baby" syndrome. A connection between nitrates and cancer is suspected by some, but has not been scientifically established.

Identifying the nitrate concentration "as nitrogen" is very important. It means that although you are specifically looking at the ion nitrate (NO_3^-), the result is being expressed as concentration of nitrogen (N). This is the common method of reporting nitrate concentrations.

When nitrates are a concern in groundwater, it is usually recommended that the two related parameters, ammonia and nitrite, be tested for also. These three compounds are intimately related. Ammonia is naturally transformed into nitrite and nitrite is quickly transformed into nitrate. Therefore, high levels of any of these three are a potential health threat.

C. HEAVY METALS

Our list of the common inorganic constituents contained the common metals iron and manganese. The *heavy metals* are less common and, as the name implies, they have a higher molecular weight. Heavy metals also tend to be greater health threats, as indicated by the MCLs listed in Table 7-2. The heavy metals which are most often a threat to groundwater quality are lead, copper, cadmium, chromium and arsenic.

Any of the heavy metals could occur naturally in groundwater, since they can be trace constituents in rocks. However, it is unusual for naturally-occurring heavy metals to reach concentrations approaching the MCLs (with the possible exception of arsenic in granitic areas). High concentrations of heavy metals in groundwater are usually, but not always, an indication of manmade contamination.

Lead and copper are a special case. It is quite possible that there is virtually no lead or copper in your groundwater source, and yet they may show up in high concentrations at your tap. The reason for this is the presence of copper piping and lead solder or, even worse, lead piping. Groundwater which is acidic may dissolve the copper and lead in your pipes. As most people know, lead is a particularly tragic contaminant for children because it can permanently affect their learning ability. Lead can also affect blood formation, kidney functioning and reproduction. The sale

Table 7-2 Heavy Metals

PARAMETER	MCL(mg/l) (Except *)	POSSIBLE SOURCE	POSSIBLE EFFECTS
Arsenic	0.05	N.O., smelters, glass, electronics waste, orchards	Skin, nervous system toxicity
Barium	2	N.O., pigments, epoxy sealants, coal	Circulatory system effects
Cadmium	0.005	N.O., galvanized pipe, batteries	Kidney effects
Chromium (Total)	0.1	N.O., mining, electroplating, pigments	Liver, kidney, circulatory disorders
Copper (at tap)	1.3*	N.O., wood preservatives, plumbing	Gastrointestinal irritation
Cyanide	0.2	Electroplating, steel, plastics, fertilizer	Thyroid, nervous system damage
Lead (at tap)	0.015*	N.O., plumbing, solder, brass alloys	Kidney, nervous system damage
Mercury	0.002	Crop runoff, batteries, electric switches	Kidney, nervous system disorders
Nickel	0.1	Metal alloys, electroplating, batteries	Heart, liver damage
Selenium	0.05	N.O., mining, smelting, coal/oil	Liver damage
Silver	0.1**	N.O., mining, photo processing	Skin discoloration, graying of eye

*= Action Level   ** = Secondary Drinking Water Standard

of lead piping and solder were banned in the U.S. in 1986. Oddly enough, there are no MCLs for lead and copper. Instead, the maximum levels set by EPA for lead and copper represent "action levels" above which some sort of treatment is required.

D. SYNTHETIC ORGANIC CHEMICALS

Synthetic organic chemicals (SOCs) are, almost exclusively, manmade chemical compounds which contain carbon atoms in their molecular structure. There are hundreds, if not thousands, of different SOCs in use today. The most common are industrial solvents and petroleum-based products. The organic chemicals which most often threaten groundwater supplies are listed in Table 7-3.

Table 7-3 Common Volatile Organic Compounds

PARAMETER	MCL (mg/l)	POSSIBLE SOURCE	POTENTIAL EFFECTS
Benzene	0.0005	Gasoline, pesticides, paint, plastic	Cancer
Carbon Tetrachloride	0.0005	Solvents	Cancer
Chlorobenzene	0.1	Solvents	Nervous system and liver effects
p-Dichlorobenzene	0.075	Room deodorants, mothballs	Cancer
1,2-Dichloroethane	0.0005	Leaded gas, fumigants, paints	Cancer
1,1 Dichloroethylene	0.007	Plastics, dyes, perfumes, paints	Cancer, liver and kidney effects
Ethybenzene	0.7	Gasoline, insecticides	Liver, kidney, nervous system
Ethylene Dibromide	0.00005	Leaded gas, soil fumigants	Cancer
Tetrachloroethylene	0.005	Solvents, dry cleaners	Cancer
Trichloroethylene	0.005	Textiles, adhesives, degreasers	Cancer
1,1,1-Trichloroethane	0.2	Adhesives, aerosols, textiles, paints, inks	Liver, nervous system effects
Vinyl Chloride	0.002	Solvent breakdown	Cancer

Many of the most common organic chemicals which find their way into groundwater are known as volatile organic compounds (VOCs). "Volatile" means that they tend to evaporate when exposed to air. Over the past 20 or 30 years, VOCs have become one of the greatest threats to groundwater quality in the world. There are several reasons for this including: 1) they are readily dissolved in groundwater, 2) they can move great distances through an aquifer, 3) they are all too often released to the environment or disposed of improperly and, 4) many of the most common VOCs are known carcinogens at extremely low concentrations. Most of the sites with the worst levels of groundwater contamination are contaminated with VOCs.

A common source of VOC contamination is leaking underground storage tanks (LUSTs). Awareness of LUSTs has increased dramatically over the last few years. But still, every year, the equivalent of a couple of Exxon Valdez oil spills leaks from underground storage tanks. Analyses of groundwater that is contaminated by gasoline (as well as some other petroleum products like fuel oil, diesel and mineral spirits) will show concentrations of benzene, toluene, ethyl-benzene and xylene - the so-called "BTEX" compounds. Among these VOCs, benzene is by far the most potentially injurious to health. Benzene is a known carcinogen with an MCL of 5 ppb.

One subgroup of VOCs which are very common groundwater contaminants are the chlorinated solvents which are widely used as degreasing chemicals in industrial facilities and by dry cleaners. These chemicals include trichloroethylene (TCE), tetrachloroethylene (or percloroethylene, PCE) and methylene chloride. These are some of the most common contaminants found at Superfund hazardous waste sites.

Another subgroup of VOCs are called trihalomethanes (THMs). It was discovered in the 1970s that THMs were sometimes produced in drinking water which had been treated with chlorine. The natural organics in the water reacted with chlorine to produce THMs. Some THMs, such as chloroform, are known carcinogens. As a result, drinking waters which are treated with chlorine are regularly tested for THMs. The production of THMs is a particular problem in surface waters with a high degree of organic material. Since groundwaters typically have very low levels of organic material, THMs are not usually a concern.

Many pesticides are also synthetic organic chemicals. Some of these, like ethylene dibromide (EDB), are also VOCs. As you might imagine, there are a wide variety of different pesticides on the market. It would take a whole book to go through them. Some of the more common ones are listed in Table 7-4. If you live in an agricultural area and are concerned about water quality, it would be in your best interest to find out which of these chemicals are in use.

Table 7-4 Common Pesticides

PARAMETER	MCL (mg/l)	POSSIBLE SOURCE	POTENTIAL EFFECTS
Alachlor	0.002	Herbicide on corn, soybeans, other crops	Cancer
Aldicarb	0.003	Insecticide on cotton, potatoes, others	Nervous system effects
Carbofuran	0.04	Soil fumigant for corn and cotton	Nervous, reproductive system effects
Chlordane	0.002	Soil treatment for termites	Cancer
2,4-D	0.07	Herbicide on wheat, corn, lawns	Liver and kidney damage
Dalapon	0.2	Herbicide on orchards, beans, lawns	Liver and kidney effects
Dioxin	0.00000003	Impurity in herbicides	Cancer
Endrin	0.002	Pesticide	Liver, kidney, heart damage
Heptachlor	0.0004	Insecticide for termites	Cancer
Lindane	0.0002	Insecticide on cattle, lumber, gardens	Liver, kidney, immune system
Methoxychlor	0.002	Insecticide for fruits, vegetables, alfalfa	Growth, liver, kidney, nerve effects

E. RADIONUCLIDES

Groundwater contamination by radioactive substances, known as
radionuclides, is almost always a result of naturally occurring radioactive
minerals in rocks. Extremely small amounts of radioactive minerals are
present in most rock types but they are often more abundant in granites.
The radionuclides of most concern in groundwater are radon, radium and
uranium. Some areas of the U.S. typically have higher levels of radionuclides than others. Radon, a gas which is formed by the decay of uranium-
238 is most often found in New England and in the Virginia-Maryland
area. High levels of radium are usually found in the Piedmont region and

the Midwest. The highest levels of uranium are found in the western mountain region.

Testing for specific radionuclides in water can be quite expensive and is often unnecessary. A more economical approach is to test waters for *gross alpha particle activity*. This test will pick up radiation from each of the radionuclides listed above. If the levels are found to be very low (it is impossible to live on this planet and not be exposed to some radiation) then it would be unnecessary to test further. The results of these analyses are given in picocuries per liter.

If significantly high levels of radioactivity are discovered in your water, then you may wish to test further for specific radionuclides.

Identifying Potential
Water Quality Hazards

Whether your drinking water comes from your own domestic well or a public water supply system, it is important to have some idea of the kind of threats to groundwater which exist in your watershed. If you obtain groundwater from your own well, a knowledge of the potential water quality threats in your community will help you determine what you should be testing for. If your water is supplied by a public water system, that water is already tested for a wide variety of water quality parameters. But testing alone cannot insure the continued quality of the water you drink. Groundwater sources need to be protected in order to prevent contamination from occurring. In order to protect groundwater from contamination, it is necessary to identify the specific water quality threats which exist within the watershed.

Protecting public groundwater sources from contamination has become a major priority for public water suppliers, state environmental agencies and the EPA. A strong emphasis on groundwater protection has arisen because of the extremely high costs and technical challenges of cleaning up groundwater once it has become contaminated. It is gradually becoming the law of the land that all communities which have public water supply wells must develop a strategy for protecting the aquifer which supplies that water.

Think back to our image of an aquifer being like a tremendous aquarium filled with sand and gravel. In this model, water supply wells would be like drinking straws sunk into the sand. Whatever gets dumped or spilled into that aquarium aquifer could end up in one of those straws.

In this chapter we will present an explanation of the most common sources of groundwater contamination and how they might impact the quality of your water source. These contaminant sources include: landfills, wastewater systems, agricultural uses, hazardous waste sites, natural contaminants and contaminants introduced through the water distribution system itself.

A. Landfills

You probably know landfills by the more colloquial title of "dump." Most public officials and landfill operators prefer the term "sanitary landfill." Up until the early 1970's, most landfills were "open burning dumps." The trash was burned out in the open, leaving behind ash and a relatively inert residue. But this practice was contributing to air pollution, so burning at landfills was eventually banned. Instead, the trash was dumped into a hole in the ground and buried. This created two new problems. The first was that landfills began to fill up quickly once the trash was no longer burned. The second, and more serious result, was that instead of pollutants going off into the air, they were now being carried down to the groundwater by infiltrating rainwater (see Figure 8-1). The rainwater trickles through the refuse and takes into solution whatever can be dissolved. This landfill "tea" is referred to as *leachate*.

The modern way to build a landfill is to place the trash within an impermeable lower liner until the landfill is filled to capacity. The rainfall that infiltrates through the trash is then collected by a leachate collection system. Once the landfill has reached its capacity, an impervious liner is placed on top so that rainwater cannot infiltrate into the landfill and make

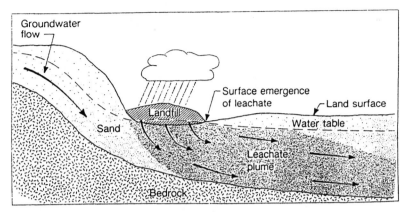

Figure 8-1 Leachate Plume Emanating From a Landfill
(U.S. Environmental Protection Agency, 1977)

more leachate. No one really knows how long one of these liners will last since they have only been in widespread use for the past two decades. Perhaps they will last fifty or a hundred years. Hopefully by that time our descendants will have figured out an inexpensive way to deal with all of those landfills. If not, we will have left them something to remember us by.

There are still many of the old style landfills without liners still around adding leachate to groundwater every day. On the rare occasions when one can see leachate at the ground surface it usually appears as a reddish and foul-smelling liquid. The chemical composition of leachate varies depending upon what was dumped into the landfill. Typically, leachate is high in most of the inorganic constituents discussed in Chapter 7. Heavy metals and VOCs are also quite common. The latter can be derived from normal household refuse, but landfills which received trash from commercial or industrial facilities tend to have higher concentrations of the more exotic chemicals. In addition, there is always the potential that hazardous wastes were inadvertently, or illegally, disposed of in a landfill.

I would include junk yards in the same general category as landfills. In junk yards, the refuse is not buried, it is temporarily stored on the ground. It is also usually composed of junked machinery which could theoretically be salvaged. But in the meantime, the same leaching process can occur.

Heavy metals, gasoline, oil and solvents can become dissolved in the groundwater. In fact this leaching process can occur at any location where refuse of any sort has been disposed of.

B. DOMESTIC WASTEWATER DISPOSAL

Included in this category of potential contaminant sources are the various methods for disposing of human wastes including *cesspools, septic systems, septage lagoons* and *wastewater treatment facilities.*

Cesspools are rather crude devices for transferring human wastes from the home to the subsurface. It is basically an underground pit. It utilizes the same level of technology as an outhouse, but at least it is buried out of sight. Fortunately, the use of cesspools as a means of waste disposal is not permitted for new housing. They have been replaced by the modern septic system.

A *septic system* is composed of a septic tank and a leaching field or pit (see Figure 8-2). This is a somewhat more sophisticated way of dealing with domestic waste. The primary treatment that happens in a septic system (if it is operating properly) is the breakdown of organic waste.

Septage lagoons are based on essentially the same technology as a cesspool, except that the so-called "lagoons" are completely open. Fortunately these lagoons are becoming rarer sights these days. They are eyesores and "nosesores" and from the point of view of impacts to groundwater, they are essentially a means of getting rid of a lot of small problems and making one large one. The small problems are the "solids" which eventually clog up any septic system or cesspool. This *septage* is pumped from clogged or failing systems and loaded into what are colloquially referred to as "honey wagons." If the community does not have access to a wastewater treatment plant which is capable of treating the septage, it may be disposed of in septage lagoons. Septage lagoons are basically pits dug in the ground which contain the solid portion of the septage but allow the water to percolate down and potentially contaminate an aquifer. Septage lagoons are usually located in areas with relatively coarse and permeable soils so that the water can drain from the septage. These are also usually the best aquifers.

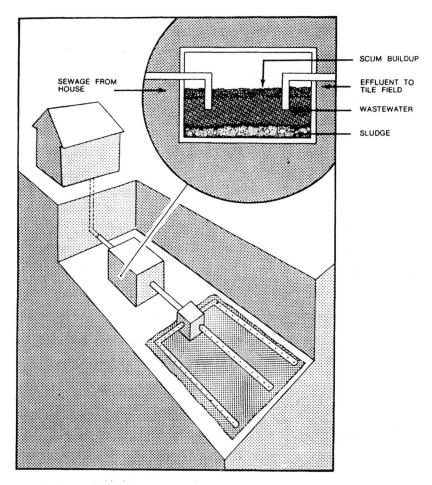

Figure 8-2 Diagram of a Domestic Septic System
(U.S. Environmental Protection Agency, 1980)

A wastewater treatment facility is designed to chemically and biologically alter wastes so that it is less noxious. The wastes are collected from each home and business via sewer lines and are collectively referred to as *sewage*. Sometimes wastewater treatment plants are also designed to treat septage. Wastewater treatment facilities come in all shapes and sizes. They could treat the waste from a small condominium or a large city.

Groundwater in very close proximity to a cesspool or septic system may contain bacteria, viruses and other biological contaminants. For this reason public and private water supply wells must be set back a certain distance from one of these wastewater disposal units (usually 100 to 400 feet).

The primary contaminant of concern from domestic wastewater disposal of all kinds is nitrate. An individual septic system introduces a relatively small amount of total nitrate into the subsurface, but the nitrate is at a relatively high concentration (roughly 30 to 40 ppm nitrate as nitrogen). As this wastewater *effluent*, as it is called, mixes with rainwater and groundwater it becomes significantly less concentrated, a process known as *dilution*. However, if there are a large number of these systems in a small area, the overall concentrations of nitrate can reach high levels. Nitrates are not only a threat to human health at concentrations above 10 ppm nitrate as nitrogen, but nitrates can also add significantly to the *eutrophication* of ponds, resulting in algae blooms and weed-choked waters. The treated effluent from a wastewater treatment plant, depending on the design, may contain nitrate as nitrogen concentrations of less than 10 ppm. Technically, the effluent may meet drinking water quality standards.

In addition to nitrates, wastewater discharge from a cesspool, septic system or wastewater treatment facility may contain just about anything that could be poured down a sink. Most people are not in the habit of pouring large quantities of toxic chemicals down the sink, but you would be surprised to know the ingredients of many household products. Some of these products contain VOCs, and even minute quantities of VOCs can be a threat to drinking water.

C. AGRICULTURAL PRACTICES

If you think about our analogy of the drinking straws in the aquarium aquifer, you would have good cause to be concerned about the water quality in an agricultural community, at least a non-organic agricultural community. Just about anything that is sprayed onto the farmers field could end up in the groundwater. The primary water quality concerns are fertilizers, animal wastes and pesticides. Both fertilizers and animal wastes are high in nitrates and it is not at all uncommon for public water supplies and domestic well waters in agricultural areas to have elevated levels of nitrates.

The uses of pesticides will vary depending primarily on the type of crop. The EPA has registered approximately 50,000 different pesticides for use in the U.S. and the rate of use and quantity of different chemicals increases every year. Since Rachel Carson's *Silent Spring* was published in the 1960's the use of DDT has been banned in many countries. Fortunately, effective replacement chemicals were found which did not have the devastating long term effects of DDT. However, these chemicals are certainly not harmless. Some of the next generation pesticides used to replace DDT have also been banned because of their toxicological effects and impacts on the environment. It is very difficult to develop chemicals which will harm only specific species of pests and not be harmful to humans.

Many pesticides are specifically formulated to biodegrade into harmless compounds within the soils so that they do not leach down into groundwaters. But if too much of these chemicals are applied or if they are applied under less than optimal conditions, groundwater contamination can still occur. An EPA survey found that the most common pesticides found in drinking water wells were dimethyl tetrachoroterepthalate (DCPA, also known as dachtal) and atrazine. The EPA classifies these chemicals as moderately toxic and slightly toxic, respectively.

Golf courses, and other land uses which rely on lush green lawns, can also cause some of the same types of groundwater contamination as agricultural uses. Intensive lawn care relies on the same types of fertilizers, pesticides and herbicides. Lawn care fertilizers used on a normal house lot can add more total nitrates than the septic system. A dense housing development with well-maintained lawns and individual septic systems could be a significant threat to the quality of the underlying groundwater.

Another potential source of high levels of ammonia and nitrates are animal wastes from large feedlots such as for cattle, pigs and fowl.

A National Pesticide Survey (NPS) was conducted by EPA's Office of Pesticides and Toxic Substances and Office of Water between 1985 and 1992 (USEPA, 1992). The survey included water quality samples from 1,300 public and domestic rural water supplies (only 10 percent were actually on farms). The NPS found nitrate concentrations above MCLs in 3.6

percent of the samples. In addition, almost 1 percent of the wells contained pesticides at levels exceeding MCLs.

D. HAZARDOUS WASTES, INDUSTRIAL CHEMICALS AND PETROLEUM PRODUCTS

The term "hazardous waste" technically refers to the waste products, or by-products, of a group of industrial chemicals which are hazardous in some way (though not necessarily toxic). However, the term is also often used as a catchall phrase for any type of industrial chemical or petroleum product which has been released to the environment. By this second definition, even gasoline which has been spilled onto the ground would be considered a hazardous waste. We will be using the term in this more general sense.

Hazardous materials, or hazardous wastes, can be considered hazardous because they are flammable, explosive, highly reactive or toxic. We will be concerning ourselves only with toxic hazardous wastes. These materials are not found solely in industrial areas. You will find these chemicals all around you unless you happen to live on an isolated mountain top.

One common source of hazardous waste contamination in groundwater is leaking underground storage tanks (*LUSTs*, or simply *USTs*). Every gasoline station has at least two or three underground storage tanks to hold the various grades of gasoline. USTs are also used for storing diesel fuel, home heating oil and industrial solvents. The major problem with USTs is that they don't last forever. They can all leak eventually. A New York Department of Environmental Conservation report estimates that 50% of steel tanks in the ground for 15 years or more are leaking (Recommended Practices for Underground Storage of Petroleum, 1984). After 20 or 30 years, most steel tanks are bound to start leaking. The technologies of tank construction and installation have improved since that 1984 study, but there are plenty of old tanks still in the ground.

The other problem is that they are "out of sight, out of mind." In the past, a leaking underground storage tank might not be discovered until it was found to be empty the day after it was filled. By then it would be too late. The contents would either be in the groundwater or well on their way. Nowadays, most USTs are required to be tested on a regular basis.

Accidents do happen. In 1984, the EPA estimated that up to a fourth of the 2.5 million gasoline storage tanks in the country could be leaking (USEPA, 1984). That same report estimates that 40% of the incidences of groundwater contamination involve gasoline.

Every industrial facility could be considered a potential source of groundwater contamination, even if there is no UST on site. This is why hazardous waste site investigations are becoming a common part of real estate transactions for industrial properties. No one wants to unknowingly buy someone else's problem. Usually a potential buyer of an industrial property will test the soils and groundwater and investigate the hazardous waste handling practices of the site they are interested in buying. The owner may not even be aware of any problems. However, liability-conscious insurance companies and banks usually require the buyer to investigate the possibility of hazardous waste contamination. An industrial facility may use chemicals which have health risks at very low concentrations and there are many ways that these chemicals could find there way to the ground and groundwater.

Some factories have floor drains that empty into the ground. If a solvent spill occurs, it is washed right into the ground. Sometimes chemicals are poured down the drain and into a septic system. In the past, chemicals were dumped into open lagoons where they would eventually seep into the groundwater. This practice is illegal in most areas, but contaminants may still remain in the ground from past lagoons.

Chemical spills happen. Drums are buried. Pipes leak. It doesn't take much of a spill to cause a relatively big problem. I know of one site where a public water supply well was contaminated with chloroform at levels above the MCLs and had to be temporarily shut down. A plume of groundwater contamination a mile long was discovered to have originated at an industrial facility. An investigation turned up the fact that several gallons of chloroform may have been poured down a drain and into a septic system. A million dollars and a few years later, the contamination plume was stopped from spreading further, but no one can say for sure when the aquifer will be clean again.

Another category of potential groundwater contaminants includes those chemicals which are applied to the ground intentionally - such as highway de-icing salts. Each year more than 11 million tons of salt are applied to roads in the United States (USEPA, 1990). Rainfall washes these salts into the groundwater. Many water supply wells located near major roads have significantly elevated levels of sodium and chloride. These chemicals obviously are not as dangerous as hazardous wastes. However, some individuals on restricted salt diets may be unknowingly consuming high levels of sodium in their drinking water.

E. NATURAL CAUSES

Not all groundwater contaminant sources are manmade. Even if you are remote from all human activities, you still need to beware of naturally occurring contaminants. In Chapter 7 we talked about the naturally-occurring inorganic chemicals such as calcium, sodium and iron. These chemicals and nutrients can be present at high levels because of natural processes.

Some rock formations and soils are enriched in heavy metals. Granites and some metamorphic rocks can have high levels of arsenic. This is a major problem in parts of New Hampshire and Massachusetts.

Some granites also contain small, but significant, amounts of uranium minerals. As the uranium decays, radon gas is formed. The gas can then enter the overlying soils and migrate upward into houses. Radon in tap water is less of a drinking water issue than an air quality one. If radon-enriched groundwater is pumped from a well it can be released into the household air as water runs from the shower or faucet. If you live in an area underlain by granite and have a domestic well, the chances are high that your water has elevated (though not necessarily dangerous) levels of radon.

Other potential contaminants that can be naturally derived from the rock types which make up aquifers include just about any metal, sulfate, chloride and fluoride.

In coastal aquifers, such as Cape Cod, it is not unusual to find low levels of naturally-occurring chloroform in the groundwater. Although the levels of chloroform in these aquifers tend be relatively low (5 to 10 ppb), they are still a potential health risk because chloroform is a carcinogenic substance.

F. WATER SYSTEM SOURCES

Some drinking water contaminants are not groundwater contaminants at all, but are introduced somewhere in the water distribution system. Two very important examples of this are the lead and copper which are sometimes leached out of water pipes and into drinking water.

Many household water pipes are made of copper. If you have copper pipes and the water in your area tends to be acidic, there is a good chance that your water will be enriched in copper. A dead give away is blue or green stains on the porcelain sink. High levels of copper can cause gastrointestinal irritation.

Although lead pipes are relatively rare, lead is a component of solder. Even the small amount of lead in a soldered pipe joint can cause elevated levels of lead in water. Lead in very small concentrations (0.05 ppm or greater) can be dangerous for small children. Lead can permanently effect the brain and can cause learning disabilities in children. Long-term exposure to lead can cause damage to the brain, kidney, nerve and reproductive systems.

Another potential contaminant in the distribution system is the group of chemicals known as trihalomethanes (THMs) which we discussed in the previous chapter. THMs can be formed when chlorine is added to water which contains natural organic compounds. Some THMs are carcinogens. This type of contamination is relatively rare in water systems which rely solely on groundwater because groundwaters are not usually chlorinated and they are usually very low in natural organic compounds. THMs formed in the manner described above are much more common in waters obtained from reservoirs and other surface water sources.

⑨

❆ ❆ ❆ ❆ ❆ ❆

Testing Your Domestic Well Water
and Interpreting the Results

L et's review what we have discussed about the issue of water quality so far. In Chapter 6 we talked about why you should want to know about the quality of the groundwater that you drink. In Chapter 7 we presented a discussion of the most common measures of water quality. In Chapter 8 we talked about the potential sources of groundwater contamination and the particular chemicals associated with these sources. Now that we have discussed the basics of water quality testing, chemistry and potential contaminant sources, it is time to consider which water quality parameters a domestic well owner might test for.

The information in this chapter can help you to make some decisions about whether to test your well water and what water quality parameters may be appropriate. However, if the concern for potential groundwater contamination of your well is great, or if you discover troubling levels of contaminants in your well, it would be advisable to enlist the aid of a knowledgeable professional (a health specialist, engineer or hydrogeologist) in this matter.

A. DECIDING WHAT TO TEST FOR

Very often well owners are under the impression that their water is of good quality because it has been "tested" by a laboratory. For them, the following statement may come as somewhat of a shock. *Just because your water has been "tested" does not mean that it does not contain harmful chemicals.* I place the word "tested" in quotation marks because one, or two, or even a few simple "tests" will not give you the whole water quality picture. That would be like a doctor checking your pulse and reflexes and then, based on that information, telling you that you are in perfect health. In order to be certain of the water quality from your well, you need to analyze the water, on a regular basis, for each hazardous chemical and microorganism which might get into your well.

Some communities have regulations requiring water quality testing for private wells. Sometimes well drillers have water samples analyzed when they complete a drinking water well for a home. Sometimes home sellers or real estate agents provide water quality testing results to prospective buyers. All of these water quality tests usually involve only a small number of water quality parameters. A typical set of water quality parameters for a tap water test might include: calcium, sodium, chloride, iron, manganese, hardness, alkalinity, pH, specific conductivity (or total dissolved solids) and bacteria. Testing for these parameters in your water will not necessarily tell you the quality of your water. It will only tell you the levels of these particular chemicals and parameters. These are not parameters that are going to be a severe health threat (except, perhaps, bacteria). These analyses might tell you if there are high "salt" levels in the water (sodium and chloride), or that the water may stain clothing (iron or manganese), or that you may have trouble making suds in the bathtub (hardness).

There is no one simple test for water quality. If you go to a doctor for a check up or because of an illness, he or she will make a thorough assessment of your health. This might include a review of medical history, a full physical exam, x-rays, blood tests, electrocardiogram, etc. Just checking your pulse and reflexes alone would certainly be a lot cheaper. But it wouldn't help very much in assessing your health or diagnosing a problem.

In the same way, testing for a few minor water quality parameters isn't going to tell you very much about the quality of your water.

The safest way to assess the quality of your well water would be to regularly analyze your tap water for the same set of parameters which are required of public water supplies. This would certainly give you a high level of comfort about the water that you drink every day. On the other hand, it would be a very costly precaution. One full round of water quality analyses might run $1,000 to $2,000. Even testing once per year would be more than most households could afford. However, if you can afford it, it would be a good idea to have a test like this done at least once on your water. Refer to Appendix C for a list of the water quality parameters which a public water system might test for.

Even if the full round of water quality parameters is beyond your household budget, you could still have your water analyzed for a set of water quality parameters which will provide you with at least some level of comfort in the quality of the water that you and your family drink. In order to come up with an abbreviated list of water quality parameters that will suit your needs, you will have to do a little bit of research. You will need to determine: 1) which water quality threats are most likely to affect your water supply and 2) what chemicals or parameters are associated with these threats. The information in this chapter will help you make this determination.

I strongly recommend that all private well owners test their water for a minimum set of water quality parameters. These parameters are discussed under the subheading of "Basic Water Quality Testing for a Domestic Well." You should also peruse the rest of the subheadings and see if any of these pertain to your situation. If they do, you should consider adding the parameters discussed in those sections to your list, if they are not included already.

BASIC WATER QUALITY TESTING FOR A DOMESTIC WELL

In spite of the earlier statement that there is no simple set of water quality tests which will ensure your safety, there is a relatively standardized set of inorganic water quality parameters that local Boards of Health and laboratories may recommend to private well owners. These are by no means a definitive list of potential harmful chemicals. In fact, most of the chemicals are not harmful at all. Some of them are even essential nutrients. Testing your water for this set of parameters will not tell you everything you need to know about the quality of your water, but it can give you an idea of how your water compares to other waters which have been similarly tested and it could provide a warning of some water quality problems. A list of these standard parameters is provided below in Table 9-1.

Table 9-1 Basic Water Quality Parameters

Alkalinity	Manganese	Lead	Sulfate
Calcium	Hardness	Sodium	Total Dissolved Solids
Chloride	Nitrate as Nitrogen	Specific Conductivity	Copper
Iron	pH	Bacteria	

If you have never had your water tested for the water quality parameters listed above, you should think about doing so. If any of the additional situations described below apply to you, I strongly urge you to test for the appropriate water quality parameters even if you decide not to test for these basic parameters. If you do go with the basic parameters, include any additional parameters described below if they seem appropriate.

IF YOU HAVE CHILDREN IN THE HOUSE

Young children are more sensitive to certain chemicals in the environment than adults are. For instance, most people are familiar with the potential threats posed by lead paint in the home. But high levels of lead are occasionally found in tap water as well. Lead is a potential problem regardless of whether your water comes from a private well or a public source. That is because the source of lead is most often from the pipes within the home.

Although lead pipes are uncommon, lead solder was often used to seal pipe joints. All lead piping and solder were banned from water supply plumbing in 1986. If your home was built before 1986, you should definitely have your water tested for lead. Until the results come back, you can reduce the risk of lead contamination by running the tap for several minutes in the morning (or any time the tap has not been used for a while) before drinking the water.

Babies and young children are also sensitive to high levels of nitrates in water. High levels of nitrate can cause a disease called methemoglobinemia. The nitrate reverts to nitrite in the body which blocks the flow of oxygen in the bloodstream. Methemoglobinemia is relatively rare in this country, but if you live in a farming or livestock area and you have young children you should definitely have your water tested for nitrate. Note that boiling water with excessive levels of nitrate does not help. In fact it only concentrates the nitrate even more.

IF YOU LIVE NEAR AN INDUSTRIAL OR COMMERCIAL AREA

Consider testing your well water for VOCs, at least once, if you live near any of the following: 1) a gasoline station or service garage, 2) industrial facilities, 3) a commercial district (such as a shopping center), 4) a dry cleaners or, 5) a landfill. The term "near" is obviously subjective. VOCs have been known to travel for miles in groundwater, but the further away you are from the source of contamination, the less likely you are to be affected by it. Strictly speaking, you would only be concerned if you were hydrologically downgradient of one of these potential contaminant sources. But unless you are quite sure that your well is not hydrologically downgradient, it would be prudent to test your water anyway. You can't assume that topographically downhill will also be hydrologically downgradient.

If your well is within one quarter mile of an industrial facility or a landfill you should add heavy metals to your list of water quality tests. Some heavy metals carry significant health risks if they are found in drinking water, particularly arsenic, cadmium, chromium and lead. A more complete list would also include mercury, selenium and silver. There are several analytical methods which a laboratory can use to detect these metals. Some are more expensive than others because they can detect lower concentrations. Be sure to request that the laboratory analyze these metals using a method which can detect down to a level at least as low as the MCL.

IF YOU LIVE IN AN AGRICULTURAL AREA

If you live in an agricultural area, consider testing for nitrates and selected pesticides. The groundwaters of almost all agricultural areas have elevated levels of nitrates. This is a relatively easy and inexpensive test to do and it is especially important if you have an infant in the house.

Pesticides are a more complicated issue. Generally pesticides today are specially formulated and applied in a manner which is intended to limit impacts to groundwater. However, there are still risks associated with pesticides. Some arise from improper storage and disposal or improper use. Sometimes pesticides which were once considered safe are later found to have potential health risks. The best way to protect yourself from these types of potential contaminants is to find out which potentially harmful chemicals are used near your home and then test your water for them. It may be helpful to check with your local Cooperative Extension Service for information on local pesticide use. The testing may be a relatively expensive proposition, depending on how many potentially harmful chemicals are in use (or were formerly in use). It may require a separate analytical procedure for each suspected chemical. You will need to evaluate your risk and weigh the pros and cons.

IF YOUR WELL IS IN GRANITE ROCK

Granite is one of the most common rock types on the planet. The water quality of most wells drilled in granite is fine. But in some cases, granites are enriched in potentially hazardous chemicals. In New England, two water quality threats in particular have been associated with some granites: arsenic and radon.

Arsenic is a heavy metal whose toxicity is well known. Check with your local Board of Health and see if wells in your area have detected traces of arsenic. The test is relatively inexpensive.

Radon is a radioactive gas which is produced in granite rocks which have low levels of uranium minerals. If there is granite rock beneath your house, consider testing both the well water and the air in the basement.

Relatively few laboratories perform routine analyses for radionuclides. You might even have to send the sample out of state. Contact your local Board of Health or the EPA for information on laboratories that can test for radionuclides. On the other hand, testing the air quality in your home for the presence of radon is relatively easy and inexpensive (around $15 to $30 dollars per test). For more information on radon testing, contact the American Cancer Society.

IF YOUR SINK BECOMES STAINED

When white porcelain sinks or tubs stain red, blue, green or black just below the faucet, it is usually because metals are coming out of solution from the tap water. This is fairly common in areas where the groundwater is on the acidic side (low pH), which includes most of the northeastern United States. In acidic groundwaters, metals can be dissolved into the water at higher concentrations. When this water finally comes into contact with the air after its long sojourn under the ground, the water becomes oxidized. Under oxidized conditions, metals tend to come out of solution and stain your sink. Red or brown stains indicate iron (basically the same thing as rust), blue or green indicates copper and black is probably manganese.

The only troubling metal in this group, in terms of health impacts, is copper. However, if you have any staining, it indicates that the waters are acidic and there may be other metals in the water which would not cause this telltale staining but which may be health hazards, metals such as lead, chromium, cadmium, selenium, arsenic, mercury. You may want to test for some or all of these. Definitely test for lead and copper.

IF YOUR WATER SOMETIMES SMELLS OR TASTES ODD

Smell and taste are very subjective measures of water quality. Most of the time your water probably does not taste like anything. Water is more or less tasteless. But it could also be that you have just gotten used to the taste. You might notice some subtle changes in the taste or smell of your water over time or you may notice that the water at friends or neighbors houses seem different. Most often taste and odor problems in drinking water are more aesthetic problems than health problems. Toxic chemicals are usually tasteless at concentrations found in groundwater. However, there may be other contaminants associated with the ones that you can taste or smell. Odor and taste problems are sometimes caused by: 1) metals (usually iron, manganese, copper or zinc) which impart a distinctive metallic taste, 2) hydrogen sulfides which cause a "rotten egg" or sulfur smell or, 3) chlorine. If your water has a metallic taste, definitely have it analyzed for lead and copper. You can also test for sulfates. Chlorine is almost always an intentional additive which may be offensive but is not harmful.

In general, if the taste or smell of your water is offensive, or if it seems to have changed over time, it would be a good idea to have your water tested thoroughly.

B. Finding a Laboratory

Finding a water quality testing laboratory is easy. Just look in the phone book under Water Testing Laboratories (or something similar). The question is how to find a good one. This could be relatively easy also. Call the lab up and talk to them about what you want to test for. Ask them to send you a brochure. You might ask your local Board of Health for a recommendation. A laboratory that does water quality analyses for a municipality or water company on a routine basis is probably a good bet. This is a fairly competitive business and prices really shouldn't be an important factor.

Also be sure that the laboratory is certified by your state to do the analyses that you need. Some laboratories are certified for only a small number of parameters. Don't be shy about asking to see copies of their certifications. And make sure they are current, certifications need to be renewed regularly.

In some regions of the country, state or county laboratories provide water quality analyses for free, or at a reduced cost, to domestic well owners. Check with your local Board of Health before paying for a private lab to do the analyses. It is unlikely that a state or county laboratory would analyze your water for all of the chemicals covered by the SDWA, so you might have to find a private lab for at least some of the analyses that you will need.

C. INTERPRETING THE RESULTS

The best way to ensure that you are correctly interpreting the water quality results from the laboratory is to have them reviewed by an experienced professional such as an environmental scientist, hydrogeologist, engineer, chemist or health professional. Someone on your local Board of Health may be willing to review the results with you. On the other hand, if you have a little knowledge of chemistry and want to learn a little bit about your water, the information below will help you to interpret the results that come back from the lab.

Each laboratory has its own way of presenting water quality analyses and sometimes they are difficult to read. A lot of information is usually presented on one page without a lot of explanation. Figure 9-1 is a typical example of a laboratory report for volatile organic compound (VOC) analysis. For each water sample there may be several forms like this for each of the different chemical types being tested for. Usually each page of a laboratory report presents the results of the parameters analyzed using the same laboratory method. In this case, the analytical method is known as EPA Method 502.2 (see column labeled "Analytical Method").

Usually the parameters being analyzed for are listed along the left side of the page. The columns to the right present information on each of the chemical parameters, the most important of which is the concentration which was detected in the water sample. In the case of the example in Figure 9-1, this column is labeled "Result." The concentrations are usually presented in milligrams per liter (mg/l) or micrograms per liter (*ug/l*). Often these units are referred to by the essentially equivalent terms parts per

MASSACHUSETTS DEP/DIVISION OF WATER SUPPLY
VOLATILE ORGANIC CONTAMINANT REPORT
(Form #7.2)

I. PWS INFORMATION:
    1. PWS ID#:   2. City/Town:
    3. PWS Name: 4. PWS Class: COM
    5. DEP Source Code/Location ID: 6. Sample Location: 7. Date Collected: 8. Collected by:
                                                                          06-18-96

    9. Is the Source Treated?   NO   10. Was the Sample Collected after Treatment?   N/A
    11. Manifolded ☐ If applicable, list the connected sources:

    _____
    12. Routine ☒   Special ☐ (explain below)
    Notes: _____
    _____

II. LABORATORY ANALYTICAL INFORMATION:
    Lab Name: Groundwater Analytical, Inc.   Lab Cert. #:   M-MA-103
    Subcontracted?   NO   Lab Sample ID#:   13611-02
    Sub. Lab Name: Sub. Lab Cert. #:
    Composited ☐ If applicable, list the composited sources:

    _____
    Notes: _____

Compound (Regulated)	Result µg/L	MCL µg/L	Detection Limit µg/L	Analytical Method	Date Analyzed
Benzene	ND	5.0	0.5	502.2	06-20-96
Carbon Tetrachloride	ND	5.0	0.5	502.2	06-20-96
1,1-Dichloroethylene	ND	7.0	0.5	502.2	06-20-96
1,2-Dichloroethane	ND	5.0	0.5	502.2	06-20-96
para-Dichlorobenzene	ND	5.0	0.5	502.2	06-20-96
Trichloroethylene	ND	5.0	0.5	502.2	06-20-96
1,1,1-Trichloroethane	ND	200.0	0.5	502.2	06-20-96
Vinyl Chloride	ND	2.0	0.5	502.2	06-20-96
Monochlorobenzene	ND	100.0	0.5	502.2	06-20-96
o-Dichlorobenzene	ND	600.0	0.5	502.2	06-20-96
trans-1,2-Dichloroethylene	ND	100.0	0.5	502.2	06-20-96
cis-1,2-Dichloroethylene	0.5	70.0	0.5	502.2	06-20-96
1,2-Dichloropropane	ND	5.0	0.5	502.2	06-20-96
Ethylbenzene	ND	700.0	0.5	502.2	06-20-96
Styrene	ND	100.0	0.5	502.2	06-20-96
Tetrachloroethylene	ND	5.0	0.5	502.2	06-20-96
Toluene	ND	1000.0	0.5	502.2	06-20-96
Xylenes (total)	ND	10000.0	0.5	502.2	06-20-96
Dichloromethane	ND	5.0	0.5	502.2	06-20-96
1,2,4-Trichlorobenzene	ND	70.0	0.5	502.2	06-20-96
1,1,2-Trichloroethane	ND	5.0	0.5	502.2	06-20-96

Figure 9-1 Example of a Laboratory Water Quality Report

million (ppm) and parts per billion (ppb), respectively. If none of the chemicals are detected, this does not necessarily mean that none are present but that if they are present they are below detectable levels. This might be denoted on the report as BDL (Below Detectable Levels) or ND (None Detected). Often there is a column of numbers which lists the lowest detectable levels for each of the chemicals, labeled "Detection Limit" in Figure 9-1. The laboratory analysis shown in Figure 9-1 also lists the MCL for each of the chemicals (note that they are listed in *ug/l*, not mg/l). For a VOC analysis of your water, you would hope that the column labeled "Result" would contain all NDs. In this case, there was a trace (0.5 *ug/l*) of cis-1,2-Dichloroethylene, which is well below the MCL.

Figuring out how to read the numbers is the easy part. Now, what do these numbers mean? How good is my water? Is it high quality, just so-so, or is it dangerous to drink? These are more difficult questions to answer. Expert opinions on what is safe and what is not safe to drink are changing all the time. Usually these determinations are based on a calculated risk, but everyone has their own tolerance for risk. Some people are devil-may-care risk takers and others are not. In general, the Maximum Contaminant Levels (MCLs) are based on some very conservative estimates of risk. Comparing the levels in your water to the MCLs is a good place to start your water quality assessment.

If the concentrations of any parameters found in your drinking water exceed the MCLs listed in Appendix B, the long-term safety of your drinking water is questionable. This does not necessarily mean that drinking a glass of this water will cause you immediate harm or that there will be serious consequences because your family has been using this contaminated water for weeks or months. Many of the MCLs, particularly the carcinogenic substances, are based on exposure over the course of a lifetime. Don't panic. Do some research. Find out what this chemical is and what the potential health impacts might be. And begin to explore options such as treating the tap water or drinking bottled water. In the vast majority of cases the water will be fine for bathing or washing. Notable exceptions to this are radon and high levels of VOCs which can evaporate into the air and potentially contaminate the household air. If in doubt about the quality of your water, check with your local Board of Health.

If none of the chemical parameters that you tested for exceed the MCLs, that is encouraging. The next level of water quality safety recognized by the EPA is the Maximum Contaminant Level Goals, or MCLGs (see Appendix B). These are stricter guidelines which are not enforceable but are recommended maximum goals for public water systems.

If the concentrations of any water quality parameters are close to, but do not exceed the MCL, then you have entered a kind of gray area. If the MCL of a particular chemical is 10 ppm and your water analysis shows 9 ppm, you should still be somewhat concerned for two reasons: 1) laboratory analyses are not always completely accurate (this is just a fact of life) and, 2) if you test your water a week later, it could easily be above 10 ppm because the concentrations of chemicals in water are constantly fluctuating. You may have taken your water sample at a time when the concentrations were on the low side. This is another reason why you should test your water on a regular basis. Among hydrogeologists and water supply engineers, the results of one water quality analysis are never considered conclusive. They almost always consider trends in water quality over time.

D. OPTIONS FOR DEALING WITH UNSATISFACTORY WATER QUALITY

The water that comes out of your tap may be unsatisfactory to you and your family for a number of reasons: 1) the levels of some chemicals exceed the MCLs and you are considered about the health effects, 2) the levels of contaminants do not exceed the MCLs but they do appear to be elevated and you are still concerned about the potential health effects or, 3) although there may not be any health impacts, the water has an objectionable taste or odor (for example, from chlorine).

The two most common responses to this situation are installing some type of water treatment system or buying bottled water for drinking. At this point we are beginning to get out of the realm of groundwater science. If you are contemplating a home treatment system or alternatives to your water source, you will need to embark on a research project which is beyond the scope of this book. We will, however, briefly review two options and provide some advice and suggestions for further reading.

BOTTLED WATER

If buying bottled water is your preferred option, just keep in mind that there is nothing magical about water in a bottle. These waters come from the same types of aquifers as public water supplies, are prone to the same types of contaminant sources and in many cases are not tested as thoroughly or as frequently as public water supplies. If you are going to be relying on bottled water as your main drinking water source and want to know more about the quality of that water, you should write to the supplier and have them provide you with copies of all water quality testing results.

HOME WATER TREATMENT

More and more, families are turning to home treatment systems, or "water purifiers," to deal with objectionable water quality. Many of these systems are reasonably priced and are quite effective for dealing with the majority of water quality problems. The difficulty lies in the fact that you cannot just buy one of these systems off the shelf, install it in your sink and then expect your problem to be taken care of forever. In fact, the proper use of these systems is a much more complicated matter than most people realize, including many of the people who sell them. If you have a home water treatment system, or are thinking about purchasing one you should be aware of a few factors which many consumers neglect to take into account. By ignoring these factors you may be just pouring money down the sink and placing your family at risk at the same time.

1. Make sure that the system you buy is capable of treating the water quality problem that you have. There is no water treatment system on the market which can take care of *every* water quality problem. Some systems work better for some chemicals than for others. You will have

to do a little research here. And don't rely solely on the advice of a salesperson. They are, on occasion, notoriously uninformed.

2. Make sure that once the system is installed you do whatever maintenance is required. There are no systems that can be simply installed and then forgotten about forever. This is every consumer's dream, but it does not exist. Filters and carbon canisters need to be replaced regularly. If they are not, in a few months you could be right back where you started, except now, you *think* you are drinking treated water.

3. If your water has radon or high levels of VOCs, treatment of your kitchen tap may not be enough. Gases will be released during showering which could be dangerous to breathe.

4. Water treatment systems are effective at removing specific contaminants but they are also effective at removing essential minerals and salts from your water as well. If your water treatment system is eliminating this source of iron, calcium, magnesium, zinc and other essential minerals from your diet, you may want to consider ways to make up the deficit.

In Appendix A of this book you will find some references to help you choose a water treatment system for your home.

10

✻✻✻✻✻✻

Protecting Groundwater Resources

Probably the most important thing that has been learned about groundwater in the past 25 years is that it is far better to protect groundwater from becoming contaminated than it is to try to clean it up once it has become contaminated. This is because the means for protecting groundwater from contamination, though often difficult to implement on a political level, are relatively simple and straightforward on a technical level. On the other hand, cleaning up groundwater once it has been contaminated is often an enormous technical challenge, is extremely expensive, and can take generations to complete.

According to a document entitled "Alternatives for Groundwater Cleanup" produced by the National Research Council (1994), there are 300,000 to 400,000 sites in the U.S. where soils or groundwater are contaminated. Estimates of cleanup costs for these sites over the next 30 years range up to *a trillion dollars*. The average cleanup cost for each of the 278 completed sites on Superfund's National Priorities List was $25 million. There are another 1,207 sites still to be cleaned up (as of December, 1996). And the list is constantly growing.

An awareness of the need for groundwater protection on the local, state and federal level arose in response to the increased incidence of widespread groundwater contamination during the 1970's and 1980's. It was discovered that even relatively small releases of substances like industrial solvents could, over time, contaminate large portions of an aquifer. It was also discovered that almost all of these releases to the groundwater could have been eliminated if these chemicals were properly stored, monitored and disposed of.

By the early to mid-1980's, local, state and federal agencies in areas with threatened groundwater resources were feverishly trying to develop methods for implementing groundwater protection measures. During this same time period, it was becoming increasingly clear to groundwater professionals and regulatory agencies that cleaning up contaminated groundwater was much more difficult than anyone imagined. In some cases, after spending millions of dollars on years of evaluations and cleanup efforts (not to mention lawyers' fees) at Superfund sites, the underlying groundwater aquifers were not significantly cleaner than when the sites were first discovered. Scientists and regulators were learning the lesson that when it comes to groundwater: an ounce of prevention is worth about a ton of cure.

We will begin this chapter by examining some of the difficulties of cleaning up contaminated groundwater. This is an important issue these days. Since the cleanup of groundwater often costs many times more than the economic worth of the resource, there are some who advocate selectively abandoning aquifers that become contaminated. The other side argues that this would let major polluters off the hook too easily and squander future potential resources.

Either way, the importance of groundwater protection is obvious. But how should this be accomplished? The EPA and virtually all states have opted for the concept of *wellhead protection*, that is, protecting the areas which contribute water to a well. There are some who argue that this approach leaves future water resources vulnerable to contamination. We will present both sides of this debate.

Since most states now have some type of wellhead protection program in place, we will discuss how these programs work, how wellhead protection areas are defined, and what type of regulations and land-use controls are used.

A. THE DIFFICULTY OF CLEANING UP CONTAMINATED GROUNDWATER

Groundwater cleanup technology is a field which has grown tremendously in the last 10 years or so. Every year there are some new techniques, or modifications of old techniques, added to the arsenal of cleanup technologies. But we have a long, long way to go. Present day groundwater cleanup technologies could be compared to 18th century medicine. The methods are crude and often of questionable efficacy. The primary difficulties in cleaning contaminated groundwater lie in: 1) the nature of groundwater occurrence and its movement, 2) the large variety of contaminants and the differences in their chemical properties, and 3) managing and funding the cleanups.

The serious contamination sources which affect rivers and air are caused by either continuous or periodic discharges of chemicals. Once these discharges are eliminated, the original quality of the river or air will return fairly rapidly. The Clean Water Act of the 1970's, which applied almost exclusively to surface waters, resulted in reduced chemical discharge to many rivers. You can now fish and swim in rivers that were previously considered health hazards. Groundwater is different. Groundwater moves very slowly and contaminants are not so readily diluted and dissipated. A one-time spill of solvents can create a *plume* of contaminated groundwater which could persist for decades or even centuries.

GROUNDWATER CONTAMINANT PLUMES

The word *plume* is used to refer to an area or, more accurately, a volume of groundwater which has become contaminated by a particular source. In most cases we are dealing with a *point source*, that is, a single and identifiable source of contamination such as a leaking underground storage tank or a chemical spill. Widespread sources of contamination, such as road salts and nitrates from lawns and septic systems, are referred to as *nonpoint sources of contamination*.

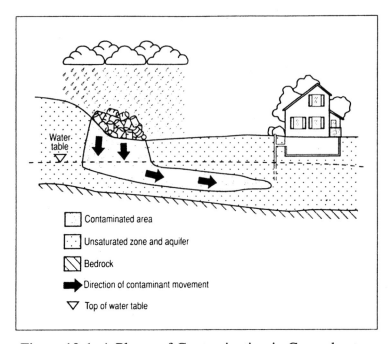

Figure 10-1 A Plume of Contamination in Groundwater
(Reprinted with permission from Home Water Treatment, published by NRAES,
Cooperative Extension, 152, Riley-Robb Hall, Ithaca, New York 14853-5701)

A typical groundwater plume is illustrated in Figure 10-1. This typical plume begins at the point where contamination was dumped, buried or spilled on the ground and extends outward in the direction of groundwater flow. As the plume extends away from the source it tends to get wider and thicker because the contamination is mixing with cleaner groundwater as it winds its way through the maze of soil particles in the aquifer. The shape of the plume and the concentration of contaminants at any given time are determined by many factors including; the rate of groundwater flow, the complexity of the aquifer, the types of contaminants and whether the source was a single spill or a constant release. Some contaminants are chemically altered as they move through the groundwater, others remain intact. Predicting the movement of plumes is a very complex task.

But let's start at the beginning. How do you even know that there is a plume of contaminated groundwater in the aquifer? Like groundwater itself, a plume cannot be directly seen. The extent of groundwater contaminant plumes are delineated primarily on the basis of water quality information derived from *monitoring wells.* Monitoring wells are small diameter (usually 2 inches) wells, often constructed of the plastic polyvinyl chloride (PVC). Typical monitoring well constructions are illustrated in Figure 10-2. Monitoring wells are installed in locations which a hydrogeologist decides can tell the most about the plume. In thick aquifers, several monitoring wells may be placed at the same location but screened at different depths. Water samples are taken from each monitoring well. These samples are then analyzed for the chemicals of concern. By plotting the concentration of chemicals on a map, the hydrogeologist can determine the extent of the plume. An example of a groundwater plume plotted in three dimensions is shown in Figure 10-3. Often it is necessary to install many wells over time and conduct numerous rounds of sampling. A groundwater plume is a moving target.

There are a growing number of technologies which can aid hydrogeologists in tracking a groundwater contaminant plume. In some cases, samples of groundwater can be obtained during the drilling process and analyzed for specific chemicals right on site. This can greatly reduce the time and effort needed to track a plume. There are also numerous "high tech" methods which do not even require wells, but these have limited applications and the results are often inconclusive. There is presently no real substitute for drilling a hole and getting a sample of the water.

GROUNDWATER CLEANUP TECHNOLOGIES

If a source of contamination is known to exist or a plume of contamination is discovered in the groundwater, the usual sequence of events is to: 1) determine the extent of the plume, 2) attempt to predict its movement in order to determine if there are any immediate threats to health or the environment (for example, downgradient water supply wells) and finally, 3) figure out a means of dealing with the groundwater contamination. Dealing with the contamination can range anywhere from ignoring it to a full scale cleanup effort.

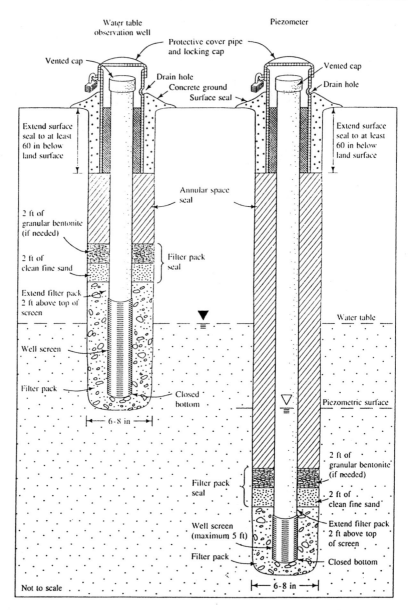

Figure 10-2 Typical Groundwater Monitoring Wells
(Wisconsin Department of Natural Resources)

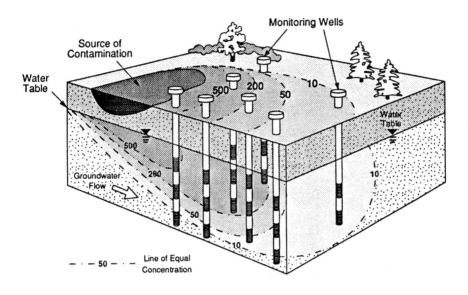

Figure 10-3 Plotting a Groundwater Contaminant Plume

So far, we have been using the word "cleanup" because it is a word that is familiar. Keep in mind that a groundwater cleanup is not like cleaning up a spill on your kitchen counter. It's more like trying to clean up a paint stain on a white carpet. You can get the bulk of the liquid off but you may never have a beautiful white carpet again. The more technical term that is used for groundwater and soil "cleanup" is *remediation*. Remediation does not imply that you will rid the aquifer of the traces of contamination, only that you will make the situation better somehow. Remediation is somewhat of a euphemism, so I will stick to the more familiar word "cleanup."

The first task in a groundwater cleanup project is to define the extent of groundwater contamination. Often this requires a considerable effort, depending on the complexity of the geology and the type of chemical contaminant. The full evaluation of a large contamination site with multiple chemical sources and complicated geology usually takes years of sustained (and costly) effort just to determine how far the contamination extends.

In most groundwater cleanups, the contaminated groundwater has to be pumped out of the aquifer in order to be treated. This is referred to as *pump and treat* technology. For all but the smallest groundwater contamination plumes this can be a major undertaking. And it can take a long time. How long? That's difficult to say with certainty since most of the major groundwater cleanup programs have not been fully completed.

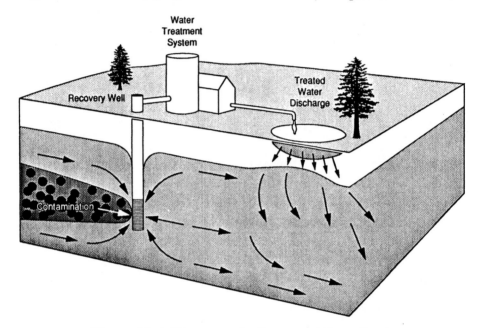

Figure 10-4 Diagram of a Pump and Treat System

Figure 10-4 illustrates a typical pump and treat scenario. The contaminated water is captured by recovery wells at the downgradient edge of the plume. This contaminated water is processed through a treatment system and then released back into the aquifer at a point downgradient of the contamination. There are many possible variations of this basic system.

Not all groundwater contaminants are treated in this way. For instance, petroleum products (such as fuel oil and gasoline) released into groundwater can sometimes be treated in the aquifer itself by introducing specific strains of petroleum-eating bacteria. This type of treatment is known as *bioremediation.* Bioremediation can be very effective for spills of petroleum products, but will not be as effective if the petroleum is mixed with other chemicals. Mixed chemicals in groundwater can be especially difficult to clean up.

Dense Non-Aqueous Phase Liquids (*DNAPLs*, pronounced "deenapples ") present another serious impediment to cleanups. DNAPL chemicals, such as many common industrial solvents, are denser than water in high concentrations and tend to sink through the aquifer. These liquids get stuck in the soil pores and sometimes accumulate in pools at the bottom of the aquifer (see Figure 10-5). Sites which are contaminated with significant quantities of DNAPL chemicals may be pumped and treated for decades (perhaps centuries) without seeing a significant improvement in groundwater quality. This is because the chemicals are dissolved into solution only a little bit at a time.

If the technical hurdles were not difficult enough, there are also political and administrative hurdles. Under ideal circumstances, the party which is responsible for the contamination should be responsible for cleaning up the groundwater. However, cleanups are seldom that simple. Who decides who is responsible and how is that decision made? Often it is technically difficult to assign guilt for a specific contamination occurrence, especially in a highly industrial area where there may be many potential sources or in a geologically complicated aquifer. Throw in some experts and a few lawyers and it can take years to decide who is responsible for cleaning up a site. In the meantime, the contaminants continue to migrate.

All of this helps to explain the seemingly poor record of federal and state groundwater cleanup efforts such as Superfund. It is not necessarily the case that the sites are mismanaged or that regulators and responsible parties are dragging their feet for financial or political reasons (although these are possible). It is more that the cleanup of groundwater contaminated with hazardous wastes is an inherently Herculean task. Sometimes it is not clear that the cleanup of a site is even possible, no matter how much time or money is spent.

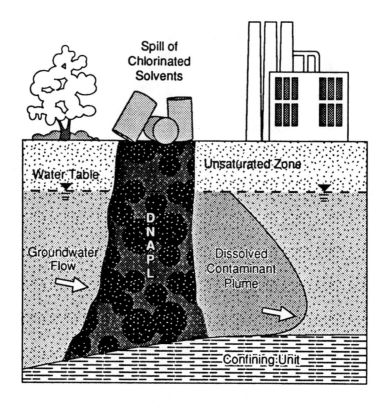

Figure 10-5 Dense Non-Aqueous Liquid in an Aquifer (DNAPL)

Groundwater contamination is like war - it is devastating, expensive and usually preventable. Clearly, our environmental capital is much better spent in the prevention of groundwater contamination.

B. GROUNDWATER PROTECTION STRATEGIES

In the 1970's, the strategies for protecting groundwater were generally referred to as "aquifer protection measures" and many municipalities and regions were developing what were referred to as "aquifer protection plans." These types of plans were relatively easy to develop. First you needed to define the aquifer area, then develop a set of protection strategies within that resource area. Aquifers are relatively easy to delineate on the basis of geology. Strategies for protecting groundwater might include restrictions on waste disposal, regulations on underground storage tanks and various types of land use controls.

The goal of aquifer protection is the protection of a vital natural resource - groundwater. The concept is simple and sound. However, the reality of aquifer protection ended up becoming a hard sell to the public. This is because almost all aquifer protection strategies involved a restriction of what property owners could do with their land. Therefore, implementation of these plans very often met resistance, particularly from large landowners, developers and industries. Areas for which land use restrictions were to be applied sometimes encompassed very large tracts of land. Some of these aquifer areas were potential water supply sources but were not presently being utilized as water supplies. The restriction of land use was viewed by many as equivalent to a government seizure of the land. Nevertheless, in the municipalities and regions where aquifer protection plans were politically feasible, they seemed to be extremely effective in protecting groundwater resources.

The federal government began to take a role in developing strategies for groundwater protection with the Amendments to the Safe Drinking Water Act (SDWA) which were passed in 1986. The SDWA set a goal for the establishment of State-run Wellhead Protection (WHP) Programs. The idea was that each state would develop its own program for wellhead protection. Wellhead protection is somewhat different from the concept of aquifer protection. For wellhead protection you protect only that portion of the aquifer which actually contributes water to a particular water supply well. That area is defined as a *Wellhead Protection Area* (WHPA).

There are two major differences between aquifer protection and wellhead protection. First of all, with wellhead protection you are only protecting the areas which are presently serving as public water supplies (wellhead protection does not apply to private wells, no matter how many there might be in an aquifer). Also, there is no provision for protecting potential future water supply resources, a shortcoming that might strike some people as singularly short-sighted. The second difference is that the effectiveness of a wellhead protection program relies on the accurate delineation of a wellhead protection area.

The wellhead protection area, also known as a *zone of contribution,* or *capture zone,* is the area of the aquifer which contributes water to the well under certain specified conditions. These specified conditions include the pumping rate of the well and the amount of recharge to the aquifer. In order to accurately delineate the wellhead protection area it is necessary to predict how groundwater will move under these conditions. Predictions are made using the mathematical formulas of groundwater flow. Most often these calculations are made by means of a computer model to solve the equations of flow. A map view and profile of a zone of contribution is shown in Figure 10-6.

Since the development of SDWA-mandated Wellhead Protection Programs, there has been very little action, on the state or local level, in the area of aquifer protection. Protecting potentially potable groundwater from contamination does not appear to be a politically feasible goal for this country in the latter part of the 20th century. As a matter of fact, protecting the water quality of existing wells is proving to be a very difficult task for many states. We will devote the remainder of this chapter to a discussion of wellhead protection: how it works (or doesn't) and how WHPA's are delineated.

C. DELINEATING WELLHEAD PROTECTION AREAS

At this point in time, almost all of the states have developed Wellhead Protection Programs and have had these programs approved by the EPA. Although the EPA had developed general guidelines for what these programs should look like, the specifics were left up to each state. As a result, each of the state wellhead protection programs are different. In fact, the implementation of these programs varies substantially from state to state. For instance, Connecticut has fairly rigorous technical requirements for the delineation of WHPA's and are attempting to require that all existing public water supply wells delineate WHPA's by a certain date. Massachusetts requires that WHPA's be delineated for all new wells, but not for existing wells. The State of Maine provides some guidance for the

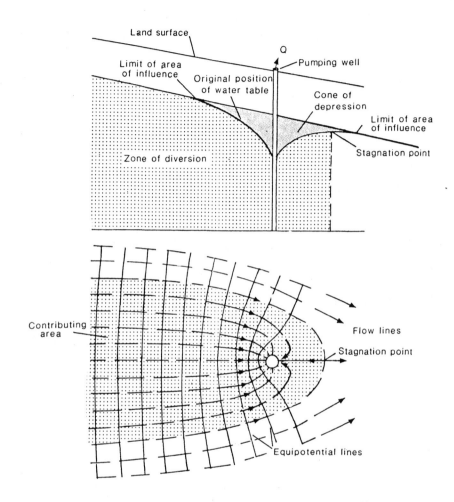

Figure 10-6 Zone of Contribution of a Well
(Morrissey, 1987)

delineation of WHPA's, encourages water suppliers to delineate them, but does not actually require them to do so.

The EPA recognizes three general methods for delineating WHPAs (USEPA, 1987,1991) including: 1) fixed radius (arbitrary and calculated),

2) analytical methods and, 3) numerical modeling. The methods are listed in order of increasing accuracy and level of effort.

FIXED RADIUS

The fixed radius method simply involves drawing a circle around the well. This circle may be either an arbitrary radius or one calculated on the basis of the relative pumping rate of the well. The primary advantages of these methods are that they are simple to do and are better than having no wellhead protection area at all. The primary disadvantage of the arbitrary fixed radius technique is that they are very crude. In reality, well capture zones are almost never circular.

Arbitrary fixed radius distances of 1,500 feet, one half mile and one mile have been used in Georgia, Massachusetts and Louisiana, respectively.

For the calculated fixed radius method, the radius of the circle is determined on the basis of the pumping rate of the well. This eliminates the "one size fits all" disadvantage. However, you still must choose a relatively arbitrary formula for determining the radius.

ANALYTICAL MODELS

Analytical models use groundwater flow equations to determine the boundaries of zones of contribution. One of the most commonly used equations is the uniform flow equation (Todd, 1980) which can be used to define the zone of contribution to a well in an aquifer with a sloping water table. Although this and other models are based on relatively simple equations, they provide a much more accurate picture of the zone of contribution than a fixed radius formula.

The primary drawbacks of the analytical models for delineating WHPAs are that numerous simplifying assumptions must be made about the aquifer and the groundwater flow in order to apply this technique. These simplifying assumptions include: 1) the permeability (transmissivity or hydraulic conductivity) is constant throughout the aquifer and that the aquifer extends infinitely in all directions, 2) the slope of the water table or potentiometric surface is constant, 3) the flow direction is uniform and 4) boundary conditions (such as rivers and aquifer boundaries) are geometrically simple. In practice, these conditions are rarely, if ever, met. Therefore, the analytical models still provide only approximate results.

NUMERICAL MODELING

Numerical modeling is the most sophisticated method of determining the zone of contribution, or WHPA, of a well. It is also the most sophisticated method for predicting groundwater flow in general and for predicting the movement of contaminants in groundwater. There are many different types of numerical groundwater models available. Some are geared toward addressing certain types of groundwater issues such as contaminant transport or three-dimensional ground-water flow.

These models are based on the same type of groundwater flow equations as the analytical models but the formulas are applied differently. Rather than assuming uniform properties and infinitely extending aquifers, as required for analytical models, numerical models can simulate more complicated aquifers because they are based on systems of grids or elements. In essence, a complicated aquifer can be divided into numerous simple pieces which are solved separately and then combined. Figure 10-7 shows an example of a model grid which was used to simulate the Martha's Vineyard aquifer in Massachusetts. Each of the boxes in the model grid can be assigned different aquifer characteristics or boundary conditions.

Figure 10-8 illustrates the significant differences in wellhead protection areas which can result from using the fixed radius, analytical modeling and numerical modeling techniques for a well site in Connecticut. The most accurate method for delineating the wellhead protection area is numerical modeling. The wellhead protection areas defined by the other techniques might result in restricting land uses in areas which do not contribute water to the well while at the same time leaving vulnerable some areas which do contribute water to the well.

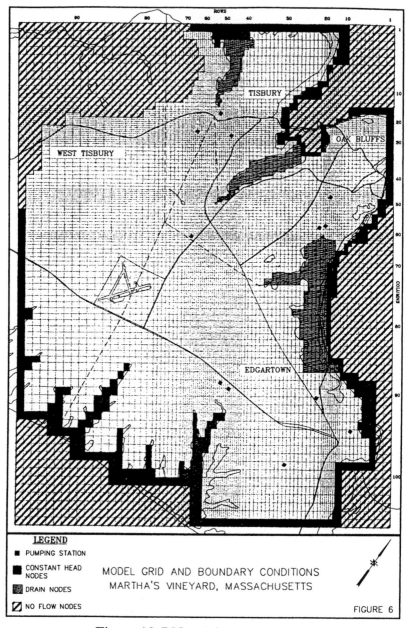

Figure 10-7 Numerical Model Grid

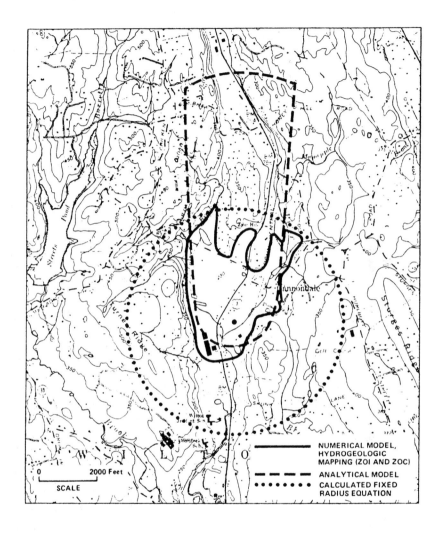

Figure 10-8 Comparison of Wellhead Protection Areas defined by
Fixed Radius, Analytical Modeling and
Numerical Modeling Techniques
(U.S. Environmental Protection Agency, 1987)

As computer hardware has become faster and able to handle larger quantities of data with greater ease, numerical groundwater flow models have been created which are capable of simulating more and more complex situations with greater accuracy. Numerical models can be used not only for defining zones of contribution to a well but for answering important questions about how aquifers work, how contaminants move through them, and how pumping wells influence each other. They can also be used to determine the relationships between groundwater and surface water bodies such as rivers and ponds. However, in order for numerical models to accurately predict groundwater movement, it is necessary to know a great deal about the aquifer and how it works. Numerical modeling is not magic. The fundamental truth which applies to all computer analyses also applies to numerical groundwater flow modeling - "Garbage in, garbage out."

D. IDENTIFYING POTENTIAL CONTAMINANT SOURCES

As was stated earlier, the WHPA of a well is the area of the earth which contributes water to that well. In general, if groundwater contaminants are going to reach a well, they will originate from somewhere within the WHPA. Therefore, once a WHPA has been defined, the next step is to identify all of the potential contaminant sources within the WHPA. This is similar to what we discussed earlier in Chapter 6 in relation to determining which water quality parameters an individual well owner should test for. In this case, we are talking about protecting a large public water supply well and the approach is more pro-active. We are not going to wait for contamination to reach the water supply well. That is the primary point of wellhead protection.

A WHPA for a municipal water supply is illustrated in Figure 10-9. This particular WHPA has numerous contamination sources including: a landfill, a gasoline station, farms, industrial facilities and a hazardous waste dump site. Once these potential sources have been identified, there are several avenues for making sure that these potential sources don't become actual sources. One possible precaution is to take steps to be sure that chemicals, particularly those in underground storage tanks, are prop-

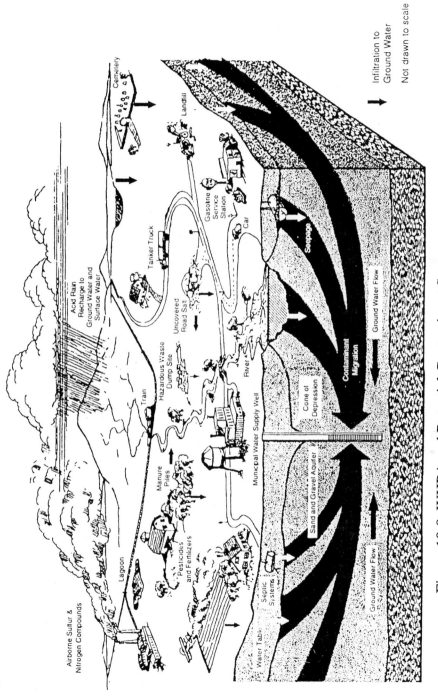

Figure 10-9 WHPA and Potential Contaminant Sources
(U.S. Environmental Protection Agency, 1993)

-erly used, transported and disposed of. It may be necessary to determine if some contamination release has already occurred.

On the other hand, there is a limit to what you can accomplish by research, inquiries and notifications. It is possible that chemical spills have occurred without anyone's knowledge. The best way to assure that the groundwater reaching the water supply is of good quality is to conduct regular groundwater monitoring at selected sites.

E. Managing Wellhead Protection Areas

Once a municipality or water company has identified a wellhead protection area and the potential contaminant sources within that area, the next step is to develop a plan to minimize the risk of contamination to that well. This is really the most difficult phase in developing a wellhead protection program. In the previous phases the issues were primarily technical ones. At this point the issues are more political and philosophical.

In accordance with the provisions of the SDWA amendments of 1986, each state has developed a statewide program for wellhead protection. This program serves as a blueprint within each state. Each state promulgates regulations, provides guidance, and sometimes even provides financial support for the implementation of wellhead protection programs. But most of the provisions for wellhead protection are actually implemented on the local level by towns and cities.

The EPA has identified several major options for wellhead management strategies. These can be broken down into regulatory and non-regulatory options.

Non-Regulatory Options

The non-regulatory options include public education, land acquisitions and groundwater monitoring. Public education can be accomplished on the federal, state or local level. The EPA, USGS and some state agencies have produced numerous documents and programs aimed at educating the public on groundwater issues in general and wellhead protection measures in particular. A partial list of these materials is included in Appendix A. Local governments and agencies also have a role in public education, although their resources are usually limited. However, the advantage of local education programs, such as municipal informational pamphlets, public forums, lectures, newspaper articles, editorials, school programs and the like, is that they can be very specific to the needs of the community.

The public acquisition of critical lands within the wellhead protection area can be an extremely effective means of protecting water quality. Many state wellhead protection programs require the acquisition of a minimum amount of land directly around a well in order to protect this area from contamination. Sometimes these lands can also be used as parks or recreation facilities.

The acquisition of large tracts of land within the wellhead protection area is also a potentially effective tool for water quality protection. However, unless the land is donated, this is generally a cost-prohibitive option for most communities. In some cases, state grants have been made available to communities for purchasing land for wellhead protection. There are also several options for negotiating conservation easements or limitations on development without actually acquiring the land.

Some form of groundwater monitoring should be a part of any wellhead protection program. The basic idea is to establish a network of groundwater monitoring points around a well so that if groundwater contamination is discovered to be migrating toward the well, steps can be taken to address the problem before the contamination affects the consumers. The monitoring wells serve as an early warning system for contamination. However, groundwater monitoring systems can never completely protect a well from contamination since there are so many potential pathways that contamination can take in an aquifer.

REGULATORY OPTIONS

Regulatory options for wellhead protection primarily consist of zoning bylaws and health regulations. Restrictions in land uses within the wellhead protection area can be instituted within the local zoning bylaws as an overlay district. Usually these are called wellhead or groundwater protection districts. Within these areas certain land uses may be prohibited, restricted or require a special permit. Prohibited land uses within a wellhead protection district might include landfills, hazardous waste disposal, storage or treatment, animal feed lots and large wastewater disposal facilities. An example of restrictions on land use would be the setting of minimum sizes for house lots in order to limit wastewater disposal within the wellhead protection area. Special permits might be required for larger wastewater systems or underground fuel storage tanks.

Health regulations can be used in conjunction with, or as an alternative to, zoning restrictions. Health regulations are administered by the local Board of Health or Health Agent. They are frequently used to control and set standards for underground storage tanks, septic systems and hazardous waste storage, transport and disposal.

The degree to which any of these controls are implemented and how effective they ultimately are in protecting groundwater resources is dependant upon: 1) local awareness of groundwater resources and the need to protect them and 2) the political climate of the state and community. Groundwater protection is, in many ways, less of a technical issue than a political one.

11

✽✽✽✽✽✽

Epilogue: The Politics of Groundwater

In the growing debate over how best to manage groundwater resources, virtually everyone involved agrees that these resources are in need of protection. The debate has primarily been about the best politically feasible means to protect these resources and who should have final authority in enacting and enforcing groundwater protection. The federal and state governments have taken the lead in this issue because they have the resources, expertise and political momentum which are necessary to work with this kind of "big picture" issue. These agencies have developed some sound technical and regulatory approaches to groundwater protection. They have also provided technical guidance and, occasionally, even funding.

But in the final analysis, the responsibility for protecting groundwater resources must fall, to a large degree, to the local communities, municipalities, counties and regional planning authorities. This is because, as we have said before, groundwater is a local resource. It is very much like the early American idea of a town common in which the citizens of a town would all share a common grazing area for cattle. The town common system depended on mutual respect, communication and cooperation. If someone abused the common, all would suffer. The town common was a shared resource among the community. Although town commons still exist in most New England towns, they are now just pleasant parks in the center of town. These communities no longer need to share grazing land. But they do need to share groundwater resources. Groundwater resources are a kind of invisible common.

The fact that this common is invisible makes it very difficult to manage. Millions of us live, literally, right on top of major groundwater aquifers without being particularly aware of what this means. Thousands of industrial firms, farms and municipalities across the country are inadvertently contributing to the pollution of aquifers (although the current regulatory environment is making this less likely). It is extremely difficult to make an issue out of something that people can't see and know little about. Hopefully, the awareness of groundwater as a vulnerable resource will begin to enter the awareness of the community as a whole through education.

Unlike the town common, the groundwater common is a resource which crosses political boundaries. This can make groundwater protection, and groundwater management in general, a truly formidable task. Why should one town enact zoning bylaws and regulations to protect the groundwater supplying wells to another town, especially when those protection measures limit land use within the community? Our tribal loyalties are not based on watersheds or groundwater basins. The extreme case of this issue is the negotiations which are going on between Israel, Jordan and the Palestinians with respect to water rights. Who will decide how the water from the Jordan River Aquifer is apportioned?

In the United States, the groundwater management conflicts we tend to encounter are a little less contentious than those in the Middle East. This is primarily because most of the U.S. has been blessed with a relative abundance of water resources. However, as the population of the country increases and as good quality groundwater resources get more difficult to find and protect, the conflicts over groundwater issues will increase. This has already been going on for some time in the arid west. These conflicts will primarily be driven by economics and by public perception. Contaminated groundwater can almost always be treated to an "acceptable" level for drinking purposes, but the question is who will pay for this treatment and who will decide what is "acceptable" quality.

Here is an example. A water supply well in Massachusetts is potentially threatened by groundwater contamination emanating from a site owned by a giant petroleum corporation. A plume of contaminated groundwater either has already reached the public water supply well or is very close. The petroleum corporation, designated the "responsible party," does not dispute this fact. The water supplier wisely decided not to pump the well once the plume was discovered because of the potential for drawing contaminated groundwater into the well. They would like the responsible party to physically contain the contamination plume before it reaches the well. This could be accomplished with a pump and treat groundwater system. However, the "responsible party" has refused to do this, stating that there is no threat to human health since the water supplier is not presently using the well.

There are some difficult questions which need to be resolved here. Should the water supplier begin to use the threatened well again in order to demonstrate that there really is a health risk which needs to be addressed? If the well is abandoned and replaced by another source, should the "responsible party" be required to share in the costs because the water supplier is reluctance to take risks with water quality?

At the same time, the state environmental authority is not obligated to take any action until the contaminant levels in the water supply well exceed the drinking water standards. Based on these regulations the petroleum corporation, or anybody else for that matter, can contaminate the water in the well up until the point when the water is no longer considered drinkable. If the water supplier wants to provide water that is completely free of carcinogenic contaminants, they would have to treat the water themselves, at their own expense. If drinking water levels are exceeded, the "responsible party" would pick up the tab for treatment but may only be responsible for treating the water to a level which would not exceed drinking water standards. The "responsible party" would not be responsible for returning the groundwater to its original quality.

So, here is our groundwater common. The town feels it has the right to uncontaminated groundwater for drinking. A petroleum giant (and many other industries, businesses and even government agencies) maintains the legal right to discharge moderate levels of chemicals into the groundwater, as long as they do not cause the water to exceed drinking water MCLs. It is apparent that we have conflicting demands on this common.

The basic dilemma behind the preceding example is this: On the one hand we have to be realistic and acknowledge that we live in an industrial society. Some level of environmental pollution seems to be inevitable and we do not want to make it economically impossible for industry to work. On the other hand, we have a diverse community which would like to utilize the groundwater common. Segments of that community have widely divergent expectations regarding the quality of drinking water. How do these conflicting uses and expectations get worked out? Often they do not. But if they get worked out at all, it is almost always accomplished on a local level, though usually under the rules of the game set down by the federal and state regulating authorities. Let's take a look at how groundwater resources are managed at these three major levels: federal, state and local.

A. THE FEDERAL ROLE IN GROUNDWATER

Every aquifer is hydrogeologically unique and is threatened by unique and identifiable potential sources of contamination. A federal agency like the EPA cannot effectively manage groundwater resources on a local, or even basin-wide, level. Instead, the EPA has been given the primary role of providing basic research into water quality issues, setting standards and enforcing compliance, particularly at major contamination sites.

There is no single, all-encompassing set of groundwater regulations which have been set by the EPA. The rules governing groundwater quality and groundwater protection come under several bodies of law. We have already mentioned the Safe Drinking Water Act (SDWA) of 1974 and the 1986 Amendments. Under this act, congress directed the EPA to set water quality standards for numerous chemicals. This list of standards is regularly updated and amended. The responsibility for enforcing these standards is generally up to each of the states. These standards are only enforced for community water systems. Individual homeowners are on their own and are not covered by these regulations. States also have the option of enforcing even stricter standards.

The SDWA also includes provisions for increased protection of groundwater, including the Sole Source Aquifer program. This program allows for the identification of aquifers which are irreplaceable sources of water supply. There is an application process whereby the EPA, individual states and even local citizens can submit an aquifer for official designation as a Sole Source Aquifer. On paper, this designation only requires that federally funded construction projects consider the potential impacts to the aquifer more carefully. In practice, a Sole Source Aquifer designation raises the awareness of most potential developers.

There are several other federal programs which deal directly or indirectly with groundwater issues. These programs are usually referred to by their acronyms: RCRA (usually pronounced "wreckra"), CERCLA and FIFRA.

The Resource Conservation and Recovery Act of 1976 (RCRA) is the EPA's primary tool for dealing with solid and hazardous waste. It includes guidelines and regulations for solid waste landfills and the production, use, storage and disposal of hazardous wastes. Many of the management and enforcement responsibilities for this program are also passed down to the states, a process referred to as "primacy." The prevention and cleanup of groundwater contamination is one of the chief concerns of this program.

The Comprehensive Environmental Response, Compensation, and Liability Act (CERCLA) of 1980 (amended in 1986) authorizes the federal government to cleanup chemical releases and hazardous waste sites which could potentially threaten human health or the environment. This program is usually referred to as "Superfund" because it allocates enormous sums of money to be used for these cleanup activities. As of December 1996 there were 1,207 sites on the National Priorities List. Most of these sites are in New Jersey, Pennsylvania and New York. Not surprisingly, most of these states have severe groundwater contamination.

The Federal Insecticide, Fungicide and Rodenticide Act (FIFRA) regulates pesticide use and marketing in the U.S. As part of the registration process for new pesticides, EPA requires that the manufacturers supply information on the leaching potential of pesticides into groundwater.

Also on the federal level there is the U.S. Geological Survey (USGS). The USGS conducts site specific geological investigations, such as geological mapping and water resources research. The Water Resources division of the USGS has been quite active over the past 20 years conducting important groundwater research such as: 1) identifying and mapping aquifers, 2) conducting water supply investigations, 3) groundwater modeling research and development, 4) basin-wide groundwater and surface water investigations and 5) research on contaminant transport.

B. THE STATES AND GROUNDWATER

Ideally, there is an environmental agency in each state which takes responsibility for seeing that the basic standards and regulations of SDWA, RCRA and FIFRA are met. At the same time, these state environmental agencies can, and often do, go much further in regulating, enforcing and implementing environmental statutes and programs which have been created by those states. The advantage of a more comprehensive state role in relating to groundwater issues is that the environmental policy makers in each state are more familiar with the issues, concerns and political climate within those states and can therefore craft more effective programs to address them. The disadvantage of this system is that the political, economic and social climates of some states are more conducive to protecting the quality of groundwater than others. That's the American home rule system.

There are so many different groundwater regulations and programs administered on the state level that it is beyond the scope of this book to discuss them. Each state has a different program and these programs are evolving relatively quickly. For more information on these programs check with the environmental agency in your state (listed in Appendix A).

C. GROUNDWATER ON THE LOCAL LEVEL

So far, the chain of command from the federal to the state level, with respect to groundwater issues, has been relatively clear-cut. If, indeed, the major responsibility for groundwater protection should be at the local level, how do we make the leap from state and federal responsibility to local control? At this point we begin to enter an uncertain and confusing landscape of authorities and responsible parties. The list of potential actors in the drama of local control of groundwater issues is extremely long and could include: municipal water suppliers or water companies, the local Board of Health, Public Works Department, Conservation Commission, Planning Board, the City or Town governing body, regional planning agency, county government, environmental groups, citizens and businesses.

The basic quandary of officials on the local level is that although they have the most to gain and to lose by decisions regarding groundwater, they have few resources do deal with these difficult, and often divisive, issues. Local governments do not usually have a lot of technical staff to help them interpret the issues related to groundwater. Often the people who make local governments work are part time or even volunteer workers. Therefore, local government decision makers need to rely on the technical expertise of others, including federal and state technical staff, hired consultants, local academic experts or regional planning commissions.

These various advisors will never have all of the answers. They are more like interpreters. It is ultimately up to the community to decide how they will manage their groundwater resources. In order to make the decisions necessary to do this, many more people in the community are going to have to learn a little bit about groundwater. There is inevitably going to have to be a greater awareness of groundwater among the people who are consumers of groundwater - the people who rely on groundwater as their primary source of drinking water. Drinking water is perhaps the most important of all consumer commodities.

It is similar in many ways to what has happened in health consciousness over the past 40 years or so. In the 1950's, people, for the most part, left it up to their family physician to worry about their health. Since then, many people have taken more responsibility for the maintenance of their own health. These days everyone seems to know a lot more about human anatomy and physiology; about cholesterol, unsaturated fats, cancer risks, the need for exercise and prenatal care. We are now much better informed consumers of food and health care. This didn't happen overnight. It was a gradual consciousness raising which was motivated by growing health concerns and facilitated by an increasingly educated public.

In order to manage our groundwater resources on a local or even regional level, we are going to have to become more educated consumers of drinking water. This education will be motivated by health concerns as well, and by the growing recognition that groundwater needs to be protected from contamination. Even in the best of times the state and federal governments could not be the sole protectors of groundwater resources. At the end of the 20th century, the political trends are decidedly more toward budget cuts and tax cuts than environmental protection. If this trend continues and environmental regulations are substantially weakened we could see groundwater contamination problems increasing at the same time as an increased demand for groundwater resources. At some point the pendulum will inevitably swing the other way, but it may be too late for aquifers that will take hundreds of years to clean up. In the future, the most decisive battles for the protection of groundwater resources are probably going to be fought on local and watershed levels.

In order to be better informed consumers of groundwater, we need to start asking questions about the quality of our water, about how our water supplies are being protected and where our groundwater comes from. This book provides the basic knowledge which is necessary to begin asking these questions and evaluating the answers. It provides a basis for citizens who are not technically trained to learn enough about groundwater to make important decisions about protecting their individual water supplies and community sources. When it comes to evaluating specific groundwater issues, this book is certainly not a substitute for the opinion of a qualified hydrogeologist, but it could provide individuals and groups with the knowledge necessary to ask the right questions and understand the advice and conclusions that are presented. Included in the Reference section are numerous books and resources for learning more about groundwater.

APPENDIX A

❀❀❀❀❀❀

Sources of Information
on Groundwater

1) Recommended Reading on Groundwater

Driscoll, F.G., 1986, *Groundwater and Wells,* Johnson Division, UOP, Inc., St. Paul, Minn.

Freeze, R.A., and J.A. Cherry, 1979, *Groundwater,* Prentice Hall, Englewood Cliffs, N.J.

Heath, R.C., 1987, Basic Groundwater Hydrology, U.S. Geological Survey Water-Supply Paper 2220.

Moore, J.E., A. Zaporozec and J. W. Mercer, 1995, *Groundwater: A Primer,* American Geological Institute, AGI Environmental Awareness Series: 1.

Todd, D.K., 1980, *Groundwater Hydrology,* John Wiley and Sons, Inc., New York, N.Y.

U. S. Geological Survey, 1976, "A Primer on Groundwater," Washington, D.C.

U.S. Environmental Protection Agency, 1987, "Handbook: Ground Water," EPA/625/6-87/016, Center for Environmental Research Information, Cincinnati, Ohio.

U.S. Environmental Protection Agency, 1990, "Ground Water, Volume 1: Ground Water and Contamination," EPA/625/6-90/016a, Center for Environmental Research Information, Cincinnati, Ohio.

2) Further Reading on Home Water Treatment

King, J., 1985, *Troubled Water,* Rodale Press, Emmaus, Pennsylvania.

Lehr, J.H., T.E. Gass, W.A. Pettyjohn and J. DeMarre, 1988, *Domestic Water Treatment,* National Water Well Association, Dublin, Ohio.

Northeast Regional Agricultural Engineering Service, 1995, *Home Water Treatment,* Cooperative Extension, Ithaca, N.Y.

3) Further Reading on Dowsing

The American Society of Dowsers, 1980, *The Water Dowsers Manual,* The American Society of Dowsers, Danville, Vermont.

Bird, C., 1979, *The Divining Hand,* E. P. Dutton, New York.

U.S. Geological Survey, 1988, "Water Dowsing," Washington: U.S. Government Printing Office.

Vogt, E.Z. and R. Hyman, 1979, *Water Witching U.S.A.,* University of Chicago Press.

4) Further Reading On Groundwater Protection

Concern, Inc., 1989, "Groundwater: A Community Action Guide," Washington, D.C. (202) 328-8160.

Conservation Foundation, 1987, *Groundwater Protection,* Washington, D.C.

Gordon, W., 1984, "A Citizen's Handbook for Groundwater Protection," Natural Resources Defense Council, New York, N.Y.

National Research Council, 1986, *Ground Water Quality Protection: State and Local Strategies,* National Academy Press, Washington D.C.

Page, W.G. (ed.), 1987, *Planning for Groundwater Protection,* Academic Press, Orlando, Florida.

Reidlich, S., 1988, "Summary of Municipal Actions for Groundwater Protection in the New England/New York Region," New England Interstate Water Pollution Control Commission, Boston, Massachusetts.

U.S. Environmental Protection Agency, 1980, "Groundwater Protection," Water Planning Division, Washington, DC.

U.S. Environmental Protection Agency, 1985, "Protection of Public Water Supplies from Groundwater Contamination," EPA/625/4-85/016, Center for Environmental Research Information, Cincinnati, Ohio.

U.S. Environmental Protection Agency, 1987, "Guidelines for Delineation of Wellhead Protection Areas," Office of Groundwater Protection.

U.S. Environmental Protection Agency, 1988, "Protecting Ground Water: Pesticides and Agricultural Practices," EPA/440/6-88-001, Office of Groundwater Protection.

U.S. Environmental Protection Agency, 1990, "Guide to Ground
    Water Supply Contingency Planning for Local and State
    Governments," EPA/440/6-90-003 (NTIS PB91-145755).

U.S. Environmental Protection Agency, 1991, "Protecting Local
    Ground Water Supplies Through Wellhead Protection,"
    EPA/570/09-91-007.

U.S. Environmental Protection Agency, 1993, "Wellhead Protection:
    A Guide for Small Communities," EPA/625/R-93/002, Center
    for Environmental Research Information, Cincinnati, Ohio.

5) Federal Agencies With Information on Groundwater

U.S. Environmental Protection Agency
Office of Ground Water and Drinking Water
U.S. Environmental Protection Agency
401 M Street, SW
Washington, DC 20460
Tel (202) 260-7593

Safe Drinking Water Hotline
(800) 426-4791

Provides information to the public on the regulations and programs
developed in response to the Safe Drinking water Act Amendments of
1986.

To order publications from the EPA's Office of Ground Water and
Drinking Water call (202) 260-7779.

EPA Regions

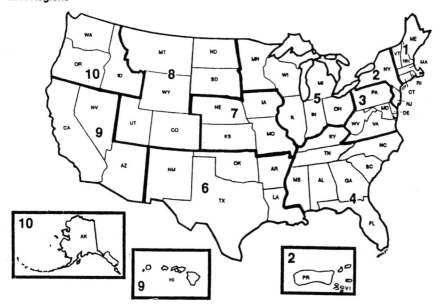

U.S. EPA Regional Offices and Groundwater Representatives

REGION 1

Office of Ground-Water
Water Management Division
U.S. EPA, Region I
JFK Federal Building
1 Congress Street
Boston, MA 02203-2211
Phone: (617) 565-3601
Fax (617) 565-4940

States Covered:
   Connecticut, Maine, Massachusetts, New Hampshire, Rhode Island, Vermont

REGION 2

Ground Water Management Section
Water Management Division
U.S. EPA, Region II
26 Federal Plaza
New York, NY 10278
Phone: (212) 264-5635
Fax: (212) 264-2194

States Covered:
New Jersey, New York, Puerto Rico, Virgin Islands

REGION 3

Office of Ground-Water
Water Management Division
U.S. EPA, Region III
841 Chestnut Street
Philadelphia, PA 19106
Phone: (215) 597-2786
Fax: (215) 597-8241

States Covered:
Delaware, District of Columbia, Maryland, Pennsylvania, Virginia, West
Virginia

REGION 4

Office of Ground-Water
Water Management Division
U.S. EPA, Region IV
345 Courtland Street, N.E.
Atlanta, GA 30365
Phone: (404) 347-3866
Fax: (404) 347-1799

States Covered:
Alabama, Florida, Georgia, Kentucky, Mississippi, North Carolina,
South Carolina, Tennessee

REGION 5

Office of Ground-Water
Water Management Division
U.S. EPA, Region V (MS-5WG-TUB9)
77 West Jackson Blvd. (WG-16J)
Chicago, IL 60604
Phone: (312) 353-1441
Fax: (312) 886-7804

States Covered:
Illinois, Indiana, Michigan, Minnesota, Ohio, Wisconsin

REGION 6

Office of Ground-Water
Water Management Division
U.S. EPA, Region VI
1445 Ross Avenue
Dallas, TX 75202-2733
Phone: (214) 655-6446
Fax: (214) 655-6490

States Covered:
Arkansas, Louisiana, New Mexico, Oklahoma, Texas

REGION 7

Office of Ground-Water
Water Management Division
U.S. EPA, Region VII
726 Minnesota Avenue
Kansas City, KS 66101
Phone: (913) 551-7745
Fax; (913) 551-7765

States Covered:
Iowa, Kansas, Missouri, Nebraska

REGION 8

Office of Ground Water
Water Management Division
U.S. EPA, Region VIII
999 18th Street
Denver, CO 80202-2405
Phone: (303) 294-1135
Fax: (303) 294-1424

States Covered:
Colorado, Montana, North Dakota, South Dakota, Utah, Wyoming

U.S. Geological Survey

U. S. Geological Survey
U.S. Department of Interior
409 National Center
Reston, VA 22092
Tel (703) 648-4000

"A Guide to Obtaining Information from the USGS"
(Circular 777) is a good resource.

Order from:
Branch of Distribution
604 S. Pickett Street
Alexandria, VA 22304

Maps are available from:

USGS Map Sales
Box 25286
Federal Center
Denver, CO 80225
(303) 236-7477

6) State Agencies With Information on Groundwater

ALABAMA
Department of
Environmental Management
Ground Water Branch
1751 W.L. Dickinson Drive
Montgomery, AL 36130
(205) 271-7773

ALASKA
Department of
Environmental Conservation
P.O. Box 0
Juneau, AK 99811-1800

Alaska Drinking Water Program
Wastewater and Water Treatment
Environmental Conservation
Department
410 Willoughby
Juneau, AK 99801
(907) 465-5316

AMERICAN SAMOA
EPA, Office of The Governor
Pago Pago, American Samoa 96799

ARIZONA
Ground Water Hydrology Section
Department of Environmental Quality
2005 N. Central Avenue
Phoenix, AZ 85004

Compliance Section
Office of Water Quality
Room 200
3033 North Central
Phoenix, AZ 85001
(602) 207-4617

ARKANSAS
Department of Health
Division of Engineering
4815 West Markham Street
Little Rock, AR 72205-3867
(501) 661-2623

Department of Pollution
Control & Ecology
P.O. Box 9583
Little Rock, AR 72219

CALIFORNIA
State Water Resources Control Board
P.O. Box 100
Sacramento, CA 95801

Division of Drinking Water and
Environmental Management
CA Department of Health Services
Room 692
714 P Street
Sacramento, CA 95814
(916) 323-6111

COLORADO
Ground Water & Standards Section
Department of Health
4210 East 11th Avenue
Denver, CO 80220

Drinking Water Program
WQCD-DW-B2
CO Department of Health
4300 Cherry Creek Drive South
Denver, CO 80222
(303) 692-3546

CONNECTICUT
Department of Environmental
Protection
Room 177, State Office Building
165 Capital Avenue
Hartford, CT 06106

Water Supplies Section
Connecticut Department of Health
Services
150 Washington Street
Hartford, CT 06106
(203) 566-1253

DELAWARE
Division of Water Resources
Ground Water Management Section
Department of Natural Resources &
Environmental Control
P.O. Box 1401
Dover, DE 19903

Public Water Systems Supervision
Program
Division of Public Health
Cooper Building
P.O. Box 637
Federal and Water Streets
Dover, DE 19903
(302) 739-5410

DISTRICT OF COLUMBIA
Department of Consumer &
Regulatory Affairs
2100 Martin Luther King Avenue
Washington, DC 20020
(202) 404-1120

FLORIDA
Department of Environmental
Regulation

Bureau of Drinking Water & Ground
Water Resources
2600 Blair Stone Road
Tallahassee, FL 32399-2400
(904) 487-1762

GEORGIA
Department of Natural Resources
Wallace State Office Building
900 East Grand Street
Des Moines, IA 50319

Drinking Water Program
Environmental Protection Division
205 Butler Street SE
Atlanta, GA 30334
(404) 651-5154

GUAM
Guam EPA
Government of Guam
130 Rojas Street/Harmon Plaza
Harmon, Guam 96911
(671) 646-8863

HAWAII
Department of Health
Ground Water Protection Program
500 Alamoana Boulevard
5 Waterfront, Suite 250
Honolulu, HI 96813

Department of Health
Environmental Management Division
P.O. Box 3378
Honolulu, HI 96801
(808) 586-4304

IDAHO
Water Quality Bureau
Division of Environmental Quality
Department of Health & Welfare

1410 North Hilton
Boise, ID 83706
(208) 334-5860

ILLINOIS
Division of Public Water Supplies
Illinois EPA
2200 Churchill Road
Springfield, IL 62794-9276
(217) 785-8653

INDIANA
Department of Environmental
Management
105 South Meridian
P.O. Box 6015
Indianapolis, IN 46206

Drinking Water Branch
Office of Water Management
100 North Senate Avenue
Indianapolis, IN 46206-6015
(317) 233-4222

IOWA
Surface & Ground Water Protection
Bureau
Department of Natural Resources
Wallace State Office Building
900 East Grand Street
Des Moines, IA 50319
(515) 281-8869

KANSAS
Department of Health and
Environment
Bureau of Water Protection
Landon State Office Building
9th Floor, 900 S.W. Jackson
Topeka, KS 66612-1290

Bureau of Water Protection
Department of Health & Environment
Building 740
Forbes Field
Topeka, KS 66620
(913) 296-5503

KENTUCKY
Division of Water
Natural Resources & Environmental
Protection Cabinet
Frankfort Office Park
18 Reilly Road
Frankfort, KY 40601

Division of Water
Drinking Water Branch
Frankfort Office Park
14 Reilly Road
Frankfort, KY 40601
(502) 564-3410

LOUISIANA
Department of Environmental Quality
P.O. Box 44066
Baton Rouge, LA 70804

Office of Public Health
Louisiana Department of Health and
Hospitals
P.O. Box 60630
New Orleans, LA 70160
(504) 568-5105

MAINE
Department of Human Services
State House Station 10
Augusta, ME 04333
(207) 287-2070

Department of Environmental
Protection
State House #17
Augusta, ME 04333

MARSHALL ISLANDS
EPA, Office of the President
Republic of Marshall Islands
Majuro, Marshall Islands 96960

MARYLAND
Department of the Environment
Room 8L
2500 Broening Highway
Dunkalk, MD 21224
(410) 631-3702

MASSACHUSETTS
Division of Water Supply
Department of Environmental Quality
Engineering
1 Winter Street
Boston, MA 02108
(617) 292-5529

Executive Office of Environmental
Affairs
100 Cambridge Street
Boston, MA 02202

MICHIGAN
Department of Public Health
P.O. Box 30035
Lansing, MI 48909

Office of Water Resources
Department of Natural Resources
P.O. Box 30028
Lansing, MI 48909

Division of Water Supply
Michigan Department of Public
Health
P.O. Box 30195
Lansing, MI 48909
(517) 335-8326

MINNESOTA
Department of Health
P.O. Box 59040
Minneapolis, MN 55459
(612) 627-5133

Pollution Control Agency
520 Lafayette Road N, 6th Floor
St. Paul, MN 55155

MISSISSIPPI
Ground Water Quality Branch
Bureau of Pollution Control
P.O. Box 10385
Jackson, MS 39289-0385

Division of Water Supply
MS Department of Health
Office U-232
P.O. Box 1700
2423 North State Street
Jackson, MS 39215-1700
(601) 960-7518

MISSOURI
Department of Natural Resources
P.O. Box 176
Jefferson City, MO 65102
(314) 751-5331

MONTANA
Water Quality Bureau
Department of Health &
Environmental Sciences
Cogswell Building, Room A206

Helena, MT 59620
(406) 444-2406

NEBRASKA
Department of Environmental Control
State House Station
P.O. Box 98922
Lincoln, NE 68509-4877

Division of Drinking Water and
Environmental Sanitation
NE Department of Health
301 Centennial Mall South
Lincoln, NE 68509
(402) 471-2541 or 0510

NEVADA
Division of Environmental Protection
201 South Fall Street, Room 221
Carson City, NV 89710

Public Health Engineering
NV Department of Human Resources
Consumer Health
505 East King Street, Room 103
Carson City, NV 89710
(702) 687-6615

NEW HAMPSHIRE
Ground Water Protection Bureau
Department of Environmental
Services
6 Hazen Drive, P.O. Box 95
Concord, NH 03302
(603) 271-3503

NEW JERSEY
Division of Water Resources
Department of Environmental
Protection
CN029
Trenton, NJ 08625-0029

Bureau of Safe Drinking Water
New Jersey Department of
Environmental Protection
P.O. Box CN-426
Trenton, NJ 06825
(609) 292-5550

NEW MEXICO
Environmental Improvement Division
New Mexico Health and Environment
Department
1190 St. Francis Drive
Santa Fe, NM 87504

Drinking Water Section
(address same as above)
(505) 827-2778

NEW YORK
Bureau of Water Quality Management
Department of Environmental
Conservation
50 Wolf Road
Albany, NY 12233-3500

Bureau of Public Water Supply
Protection
New York Department of Health
2 University Place, Room 406
Albany, NY 12203-3313
(518) 458-6731

NORTH CAROLINA
Ground Water Section
Department of Environment, Health &
Natural Resources
P.O. Box 27687
Raleigh, NC 27611-7687
(919) 733-2321

NORTH DAKOTA
Division of Water Supply and
Pollution Control
Department of Health
1200 Missouri Avenue
Bismarck, ND 58502
(701) 221-5225

NORTHERN MARIANA ISLANDS
Division of Environmental Quality
P.O. Box 1304
Saipan, Mariana 96950
(607) 234-6114

OHIO
Division of Ground Water
Ohio Environmental Protection
Agency
Box 1049
1800 Watermark Drive
Columbus, OH 43266-0149
(614) 644-2752

OKLAHOMA
Department of Pollution Control
P.O. Box 53504
Oklahoma City, OK 73152

Water Quality Programs
Department of Environmental Quality
1000 NE Tenth Street
Oklahoma City, OK 73117
(405) 271-5205

OREGON
Department of Environmental Quality
811 SW 6th Avenue
Portland, OR 97204-1334

Drinking Water Program
Health Division
Department of Human Resources

800 Northeast Oregon Street
Portland, OR 97214-0450
(503) 731-4010

PALAU
Palau Environmental Protection
Board
Republic of Palau
P.O. Box 100
Koror, Palau 96940

PENNSYLVANIA
Office of Environmental Management
Department of Environmental
Resources
P.O. Box 2063
Harrisburg, PA 17120

Division of Water Supplies
Department of Environmental
Resources
P.O. Box 2357
Harrisburg, PA 17120

Division of Drinking Water
Management
Department of Environmental
Resources
P.O. Box 8467
Harrisburg, PA 17107
(717) 787-9035

PUERTO RICO
Water Quality Area
Environmental Quality Board
Box 11488
Santurce, PR 00910

Water Supply Supervision Program
Puerto Rico Department of Health
P.O. Box 70184
San Juan, PR 00936
(809) 754-6010

RHODE ISLAND
Department of Environmental
Management
9 Hayes Street
Providence, RI 02903

Division of Drinking Water Quality
Rhode Island Department of Health
75 Davis Street, Cannon Building
Providence, RI 02908
(401) 277-6867

SOUTH CAROLINA
Bureau of Water Supply & Special
Programs
Department of Health and
            Environmental Control
2600 Bull Street
Columbia, SC 29201
(803) 734-5310

SOUTH DAKOTA
Division of Environmental Regulation
Department of Water and Natural
Resources
Joe Foss Building
Pierre, SD 57501-3181

Office of Drinking Water
523 Capital Avenue
Pierre, SD 57501
(605) 773-3754

TENNESSEE
Department of Health and
Environment
Division of Water Supply
150 Ninth Avenue, North
Nashville, TN 37219-5404

Division of Water Supply
Tennessee Department of
Environment & Conservation

401 Church Street
Nashville, TN 37243-1549
(615) 532-0191

TEXAS
Texas Department of Health
1100 West 49th Street
Austin, TX 78756

Texas Water Commission
P.O. Box 13087
Austin, TX 78711-3087

Water Utilities Division
Natural Resource Conservation
Commission
(address same as above)
(512) 908-6930

UTAH
Bureau of Drinking Water/Sanitation
Division of Environmental Health
288 North 1460 West
Salt Lake City, UT 84116-0690

Bureau of Water Pollution Control
Division of Environmental Health
288 North 1460 West
Salt Lake City, UT 84114-0700

Division of Drinking Water
Utah Department of Environmental
Quality
(address same as above)
(801) 538-6159

VERMONT
Division of Environmental Health
Department of Health
60 Main Street
Burlington, VT 05401

Agency of Natural Resources
1 South Building
103 Main Street
Waterbury, VT 05676
Water Supply Program
Vermont Department of
Environmental Conservation
103 South Main Street
Waterbury, VT 05671
(802) 241-3400

VIRGINIA
Water Control Board
P.O. Box 11143
Richmond, VA 23230-1143

Division of Water Supply Engineering
Virginia Department of Health
1500 East Main Street
Richmond, VA 23219
(804) 786-1766

VIRGIN ISLANDS
Department of Planning & Natural
Resources
179 Altona & Welgunst
St. Thomas, VI 00820
Planning and Natural Resources
Government of Virgin Islands
Nifky Center
Room 231
St. Thomas, VI 00802
(809) 774-3320

WASHINGTON
Department of Social and Health
Services
Olympia, WA 98504

Department of Ecology
Mail Stop PV 11
Olympia, WA 98504

Drinking Water Division
Department of Health
Airdustrial Center
Building #3
P.O. Box 47822
Olympia, WA 98504-7822
(206) 753-1280

WEST VIRGINIA
Office of Environmental Health
Services
815 Quarrier Street, Suite 418
Charleston, WV 25305

Department of Natural Resources
1800 Washington Street, East
Charleston, WV 25305

WISCONSIN
Division of Environmental Standards
Department of Natural Resources
P.O. Box 7921
Madison, WI 53707
(608) 267-7651

WYOMING
Department of
Environmental Quality
Water Quality Division
Herschler Building, 4th Floor
122 West 25th
Cheyenne, WY 82002
(307) 777-7781

7) Sources of Information on Groundwater in Canada

ALBERTA
Head of Hydrogeology Branch
Earth Sciences Division
Alberta Environment
14th Floor, Standard Life Center
10405 Jasper Avenue
Edmonton, Alberta T5J 3N4

BRITISH COLUMBIA
Director of Water
Management Branch
Planning and Resources
Management Division
Department of the Environment
Parliament Buildings
Victoria, B.C. V8V 1X5

MANITOBA
Water Resources Branch
Department of Natural Resources
1577 Dublin Avenue
Winnipeg, Manitoba R3T 3J5

NEW BRUNSWICK
Director of Water Resources Branch
Department of the Environment
P.O. Box 6000
Fredericton, N.B. E3B 5H1
Newfoundland
Director of Water Resources Division
Department of the Environment
P.O. Box 4750
St. Johns, Newfoundland A1C 5T7

NOVA SCOTIA
Director of Environmental Asessment
Division
Department of the Environment
P.O. Box 2107
Halifax, Nova Scotia B3J 3B7

ONTARIO
Director, Water Resources Branch
Environmental Planning Division
Ontario Environment
135 St. Clair Avenue W.
Toronto, Ontario M4V 1P5

PRINCE EDWARD ISLAND
Head of Water Resources
Technical Services Division
Department of Community and
Cultural Affairs
P.O. Box 2000
Charlottetown, P.E.I. C1A 7H8

QUEBEC
Directeur des Eaux Souterraines
Direction Generale des Inventaires et
de la Recherche Environment
194 Avenue Saint-Sacrement
Quebec, Quebec G1N 4J5
Saskatchewan
Manager
Saskatchewan Water Corporation
2121 Saskatchewan Drive
Regina, Saskatchewan S4p 3Y2

8) Other Sources Of Information

American Planning Association
(Headquarters)
1776 Massachusetts Avenue, NW
Washington, DC 20036
(202) 872-0611

American Planning Association
Research Department (Technical
Support)
1313 E. 60th Street
Chicago, IL 60637
(312) 955-9100

American Society of Civil Engineers
345 E. 47th Street
New York, NY 10017-2398
(212) 705-7496
(800) 548-ASCE

American Water Works Association
6666 West Quincy Avenue
Denver, CO 80235
(303) 794-7711

National Ground Water Association
6375 Riverside Drive
Dublin, OH 43017
(800) 551-7379

National Rural Water Association
P.O. Box 1428
2915 South 13th Street
Duncan, OK 73534
(405) 252-0629
(Also see list of Rural Water State
Associations below)
Rural Water State Associations

Alabama Rural Water Association
4556 South Court Street

Montgomery, AL 361 OS
(205) 284-1489

Arizona Small Utilities Association
1955 W. Grant Road, Suite 125
Tucson, AZ 85745
(602) 620-0230

Arkansas Rural Water Association
PO. Box 192118
Little Rock, AR 72219
(501) 568-5252

California Rural Water Association
216 W. Perkins Street, Suite 204
Ukiah, CA 95482
(707) 462-1730

Colorado Rural Water Association
2648 Santa Fe Drive, #10
Pueblo, CO 81006
(719) 545-6748

Connecticut & Rhode Island Rural
Water Association
11 Richmond Lane
Willimantic, CT 06226-3825
(203) 423-6737

Delaware Rural Water Association
P.O. Box 118
Harrington, DE 19952-0118
(302) 398-9633

Florida Rural Water Association
1391 Timberlane Road, Suite 104
Tallahassee, FL 32312

Georgia Rural Water Association
P.O. Box 383

Barnesville, GA 30204
(404) 358-0221

Idaho Rural Water Association
P.O. Box 303
Lewiston, ID 83501
(208) 743-6142

Illinois Rural Water Association
401 South Vine
Mt. Pulaski, IL 62548
(217) 792-5011
Indiana Water Association
P.O. Box 103
Sellersburg, IN 47172
(812) 246-4148

Iowa Rural Water Association
1300 S.E. Cummins Road, Suite 103
Des Moines, IA 50315
(515) 287-1765

Kansas Rural Water Association
P.O. Box 226
Seneca, KS 66538
(913) 336-3760

Kentucky Rural Water Association
P.O. Box 1424
Bowling Green, KY 42102-1424
(502) 843-2291

Louisiana Rural Water Association
P.O. Box 180
Kinder, LA 70648
(318) 738-2896

Maine Rural Water Association
14 Maine Street, Suite 407
Brunswick, ME 04011
(207) 729-6569

Maryland Rural Water Association
P.O. Box 207
Delmar, MD 21875
Salisbury, MD 21801
(301) 749-9474

Michigan Rural Water Association
P.O. Box 17
Auburn, Ml 48611
(517) 662-2655

Minnesota Rural Water Association
P.O. Box 1995
Hattiesburg, MS 39403-1995
(601) 544-2735

Missouri Rural Water Association
P.O. Box 309
Grandview, MO 64030
(816) 966-1522

Montana Rural Water Systems
Association
925 7th Avenue South
Great Falls, MT 59405
(406) 454-1151

Nebraska Rural Water Association
P.O. Box 837
Overton, NV 89040
(702) 397-8985

New Jersey Association of Rural
Water & Wastewater Utilities
703 Mill Creek Road, Suite D4
Manahawkin, NJ 08050
(609) 597-4000

New Mexico Rural Water Users
Association
3218 Silver, SE
Albuquerque, NM 87106
(505) 255-2242

New York State Rural Water
Association
P.O. Box 487
Claverack, NY 12513
(518) 851-7644

North Carolina Rural Water
Association
P.O. Box 540
Welcome, NC 27374

North Dakota Rural Water Systems
Association
Route 1, Box 34C
Bismarck, ND 58501

Northeast Rural Water Association
512 St. George Road
Williston, VT 05495
(802) 878-3276

Ohio Association of Rural Water
Systems
P.O. Box 397
Grove City, OH 43123
(614) 871-2725

Oklahoma Rural Water Association
1410 Southeast 15th
Oklahoma City, OK 73129
(405) 672-8925

Oregon Association of Water Utilities
1290 Capitol Street, NE
Salem, OR 97303
(503) 364-8269

Pennsylvania Rural Water Association
138 West Bishop Street
Bellefonte, PA 16823
(814) 353-9302

South Carolina Rural Water
Association
P.O. Box 479
Clinton, SC 29325
(813) 833-5566

South Dakota of Rural Water Systems
5009 West 125th Street, Suite 5
Sioux Falls, SD 57106

Tennessee Association of Utility
Districts
P.O. Box 2529
Murfreesboro, TN 37133-2529
(615) 896-9022

Texas Rural Water Association
1616 Rio Grande Street
Austin, TX 78701
(512) 472-8591

Rural Water Association of Utah
P.O. Box 661
Spanish Fork, Utah 84660
(801) 798-3518

Virginia Rural Water Association
133 West 21st Street
Buena Vista, VA 24416
(703) 261-7178

Washington Rural Water Association
P.O. Box 141588
Spokane, WA 99214-1588
(509) 924-5568

West Virginia Rural Water
Association
P.O. Box 225
Teays, WV 25569
(304) 757-0985

EPA Drinking Water Standards[*]

Abbreviations

MCLG -   Maximum Contaminant Level Goal. A non-enforceable concentration of a drinking water contaminant that is protective of adverse human health effects and allows an adequate margin of safety.

MCL -   Maximum Contaminant Level. Maximum permissible level of a contaminant in water which is delivered to any user of a public water system.

F -   Final Regulation

D -   Draft Regulation

L -   Listed for Regulation

P -   Proposed Regulation

T -   Tentative Regulation (not officially proposed)

NA -   Not Applicable

PS -   Performance Standard 0.5 NTU - 1.0 NTU

TT -   Treatment Technique

---

[*] Excerpted Drinking Water Regulations and Health Advisories, USEPA Office of Water, EPA 822-R-96-001, JULY 1996.

CANCER GROUPS

Group A: Human Carcinogen
Sufficient evidence in epidemiologic studies to support causal
association between exposure and cancer.

Group B: Probable Human Carcinogen
Limited evidence in epidemiologic studies and/or sufficient evidence
from animal studies to support causal association between exposure
and cancer.

Group C: Possible Human Carcinogen
Limited evidence from animal studies and inadequate or no data in
humans to support causal association between exposure and cancer.

Group D: Not Classifiable
Inadequate or no human and animal evidence of carcinogenicity.

Group E: No Evidence of Carcinogenicity for Humans
No evidence of carcinogenicity in at least two adequate animal tests in
different species or inadequate epidemiologic and animal studies.

Table B-1
Drinking Water Standards and Health Advisories

Chemicals	Status Reg.	MCLG (mg/l)	MCL (mg/l)	Cancer Group
		Standards		
Organics				
Aciflourfen	T	zero	NA	B2
Acrylamide	F	zero	TT	B2
Acrylonitrile	T	zero	NA	B1
Adipate (diethylhexyl)	F	0.40	0.40	C
Alachlor	F	zero	0.002	B2

Table B-1 (Continued)

Chemicals	Status Reg.	MCLG (mg/l)	MCL (mg/l)	Cancer Group
Aldicarb	D	0.007	0.007	D
Aldicarb sulfone	D	0.007	0.007	D
Aldicarb sulfoxide	D	0.007	0.007	D
Atrazine	F	0.003	0.003	C
Bentazon	T	0.02	NA	D
Benzene	F	zero	0.005	A
Benzo(a)pyrene (PAH)	F	zero	0.002	B2
Bromodichloromethane (THM)	P	zero	0.01	B2
Bromoform (THM)	P	zero	0.01	B2
Carbofuran	F	0.040	0.040	E
Carbon tetrachloride	F	zero	0.005	B2
Chloral hydrate	P	0.040	0.060	C
Chlordane	F	zero	0.002	B2
Chlorodibromomethane (THM)	P	0.060	0.100	C
Chloroform (THM)	P	zero	0.1	B2
Cyanazine	T	0.001	NA	C
2,4-D	F	0.07	0.07	D
Dalapon	F	0.02	0.02	D
Di[2-ethylhexyl]apidate	F	0.40	0.40	C
Dibromochloropropane (DBCP)	F	zero	0.0002	B2
Dichloroacetic acid	P	zero	0.06	B2
Dichlorobenzene o-	F	0.60	0.60	D
Dichlorobenzene m-	F	0.60	0.60	D
Dichlorobenzene p-	F	0.075	0.075	C
Dichloroethane (1,2-)	F	zero	0.005	B2
Dichloroethylene (1,1-)	F	0.007	0.007	C
Dichloroethylene (cis-1,2-)	F	0.07	0.07	D
Dichloroethylene (trans-1,2-)	F	0.10	0.10	D
Dichloromethane	F	zero	0.005	B2
Dichloropropane (1,2-)	F	zero	0.01	B2

Table B-1 (Continued)

Chemicals	Status Reg.	MCLG (mg/l)	MCL (mg/l)	Cancer Group
Dichloropropene (1,3-)	T	zero	NA	B2
Diethylhexyl phthalate (PAE)	F	zero	0.01	B2
Dinoseb	F	0.01	0.01	D
Diquat	F	0.02	0.02	D
Endothall	F	0.10	0.10	D
Endrin	F	0.002	0.002	D
Epichlorohydrin	F	zero	TT	B2
Ethylbenzene	F	0.70	0.70	D
Ethylene dibromide (EDB)	F	zero	0.0001	B2
Gylphosphate	F	0.70	0.70	E
Heptachlor	F	zero	0.0004	B2
Heptachlor epoxide	F	zero	0.0002	B2
Hexachlorobenzene	F	zero	0.001	B2
Hexachlorobutadiene	T	0.001	NA	C
Hexachlorocyclopentadiene	F	0.05	0.05	D
Lindane	F	0.0002	0.0002	C
Methoxychlor	F	0.04	0.04	D
Monochlorobenzene	F	0.10	0.10	D
Oxamyl (Vydate)	F	0.20	0.20	E
Pentachlorophenol	F	zero	0.00	B2
Picloram	F	0.50	0.50	D
Polychlorinated biphenyls (PCBs)	F	zero	0.001	B2
Simazine	F	0.004	0.004	C
Styrene	F	0.10	0.10	C
2,3,7,8-TCDD (Dioxin)	F	zero	3.00e-08	B2
Tetrachloroethylene	F	zero	0.01	-
Toluene	F	1.00	1.00	D
Toxaphene	F	zero	0.003	B2
2,4,5-TP	F	0.05	0.05	D
Trichloroacetic acid	P	0.30	0.06	C
Trichlorobenzene (1,2,4-)	F	0.07	0.07	D
Trichloroethane (1,1,1-)	F	0.20	0.20	D

Table B-1 (Continued)

Chemicals	Status Reg.	MCLG (mg/l)	MCL (mg/l)	Cancer Group
Trichloroethane (1,1,2-)	F	0.003	0.005	C
Trichloroethylene	F	zero	0.01	B2
Vinyl chloride	F	zero	0.00	A
Xylenes	F	10.00	10.00	D
INORGANICS				
Antimony	F	0.01	0.01	D
Arsenic	-	-	0.05	A
Asbestos (fibers/l > 10um length)	F	7 MFL	7 MFL	A
Barium	F	2.00	2.00	D
Beryllium	F	0.00	0.04	B2
Bromate	L	zero	0.01	-
Cadmium	F	0.01	0.01	D
Chloramine	P	4.00	4.00	-
Chlorine	P	4.00	4.00	D
Chorine dioxide	T	0.30	0.80	D
Chlorite	L	0.08	1.00	D
Chromium (total)	F	0.10	0.10	D
Copper (at tap)	F	1.30	TT	D
Cyanide	P	0.20	0.20	D
Flouride	F	4.00	4.00	-
Lead (at tap)	F	zero	TT	B2
Mercury (inorganic)	F	0.00	0.00	D
Nickel	F	0.10	0.10	D
Nitrate (as N)	F	10.00	10.00	-
Nitrite (as N)	F	1.00	1.00	-
Nitrate + Nitrite (both as N)	F	10.00	10.00	-
Selenium	F	0.05	0.05	-
Sulfate	P	500	500	-
Thallium	F	0.001	0.002	-

Table B-1 (Continued)

RADIONUCLIDES	Status Reg.	MCLG (mg/l)	MCL (mg/l)	Cancer Group
Beta particle and photon activity	P	zero	4mrem	A
Gross alpha particle activity	P	zero	15 pCi/L	A
Radium 226	P	zero	20 pCi/L	A
Radium 228	P	zero	20 pCi/L	A
Radon	P	zero	300 pCi/L	A
Uranium	P	zero	20 ug/L	A

APPENDIX C

❀ ❀ ❀ ❀ ❀ ❀

Drinking Water Monitoring
for Community Groundwater Sources[*]

TABLE C-1

CONTAMINANT	SAMPLE FREQUENCY	APPROXIMATE COST
Inorganics[1]	Once Every 3 Years	$225
Nitrate	Annually	$20
Nitrite	Once Every 3 Years	$20
Lead and Copper	Once or Twice Per Year	$40
Coliform Bacteria	Quarterly (every 3 months)	$25
Radiation (Gross Alpha)	Quarterly for 4 Years	$80
Volatile Organics[2]	Annual	$200
Synthetic Organics[3]	Quarterly for 3 Years	$200 to $2,000
Sulfate	Once Every 3 Years	$20
Secondary Contaminants[4]	Recommended Annually	$300

[1] Inorganic Compounds consist of the following: antimony, arsenic, barium, berylium, cadmium, chromium, cyanide, fluoride, lead, mercury, nickel, nitrate, selenium, sodium, silver and thallium.

[2] Volatile Organic Compounds to be analyzed using EPA Method 502.2 or 524.

[3] Synthetic Organic Compounds: Primarilly pesticides. Price depends on which SOCs are analyzed for.

[4] Secondary Contaminants consist of the following: aluminum, chloride, color, copper, corrosivity, fluoride, foaming agents (MBAS), iron, manganese, odor, pH, silver, sulfate, Total Dissolved Solids and zinc.

---

[*] Excerpted COMMONWEALTH OF MASSACHUSETTS, DEPARTMENT OF ENVIRONMENTAL PROTECTION, JULY, 1996

Glossary of Groundwater Terms

Acidic: The property of a compound that reacts with a base to form a salt. Having a pH greater than 7.

Analytical Model: A means of predicting groundwater flow based on analytical equations, with or without a computer. Analytical models require numerous simplifying assumptions. For instance, the aquifer is assumed to have the same characteristics in all directions for an infinite distance. Analytical models are usually simpler to apply than **numerical models** but cannot accurately simulate complicated aquifer geometries.

Alluvial: Deposited by a stream or other running water.

Alluvium: A general term for the unconsolidated materials deposited by a stream or other running water.

Anion: A negative **ion.**

Aquiclude: A geologic formation which limits the movement of groundwater (e.g., clay, silt or shale). Same as **confining unit.**

Aquifer: A water-bearing geologic formation, or part of a formation, capable of yielding significant quantities of water to wells, springs or rivers.

Aquitard: A **confining unit** which is not completely confining but allows some water to penetrate. See **semi-confining unit.**

Artesian Aquifer: 1) A confined aquifer in which water from a well could naturally flow, 2) any confined aquifer or sometimes, 3) a bedrock aquifer.

Artesian Well: 1) A well in an artesian aquifer from which water flows without pumping. 2) A bedrock well.

Basic: The property of a compound which reacts with an acid to form a salt. Having a pH greater than 7.

Basin (Groundwater): The area of an aquifer which contributes groundwater to a common point or area of discharge (such as a river or lake).

Basin (Surface Water): The area of land which contributes surface runoff to a common point or area (such as a river or lake).

Bedrock: A mass of rock which is not underlain by anything except more rock.

Bedrock Aquifer: An aquifer in bedrock, as opposed to an **unconsolidated aquifer.**

Bioremediation: A method for treating groundwater and soils contaminated with certain organic compounds (such as petroleum products) using bacteria to break down the contaminants.

Boundary: In hydrogeology, the term boundary is usually synonymous with confining unit. Sometimes, particularly in groundwater computer modeling, reference is made to positive and negative boundaries. A positive, or constant head, boundary would be a water body whose head is considered to be constant. Confining units would be considered no-flow, or negative, boundaries.

BTEX: Collectively, benzene, toluene, ethyl benzene and xylenes, all compounds found in gasoline.

Capillarity: The force which holds water in soils at pressure heads which are less than atmospheric pressure. The same principle which allows water in a straw to rise slightly above the water surface without any outside force.

Capillary Fringe: The zone above the water table which is saturated with water but at pressure heads which are less than atmospheric pressure. Although this zone is saturated, the level of water in a well screened across the water table will be at the water table. The water held in the capillary fringe are held there by **capillarity.**

Capillary Zone: Same as **capillary fringe.**

Capture Zone: The land area which contributes recharge to a well under a specific set of circumstances. Often used to define a Wellhead Protection Area. Same as **zone of contribution.**

Cation: A positive *ion.*

CERCLA: The Comprehensive Environmental Response, Compensation and Liability Act, also referred to as *Superfund.*

Cesspool: An on-site wastewater disposal system which consists simply of a tank or pit for transferring wastewater to the ground.

Cone of Depression: The area around a well in which water levels are lowered during pumping of the well.

Confined Aquifer: An aquifer bounded above and below by confining units and in which the groundwater is at least partially isolated from the atmosphere.

Confining Unit (or bed): A geologic formation which limits the movement of groundwater (e.g., clay, silt or shale).

Contaminant Plume: A relatively discrete body of contaminated groundwater originating from a specific source, usually elongated in the direction of groundwater flow.

Contours: Lines on a map connecting points of equal elevation. Water table or potentiometric contours can be used to define groundwater gradients and flow directions.

Dense Non-Aqueous Phase Liquid: A liquid that is denser than water and does not readily dissolved in water. Abbreviated as *DNAPL.*

Dilution: The process where chemicals in solution become less concentrated by mixing with waters containing lower concentrations of those chemicals.

DNAPL: Dense non-aqueous phase liquid. A liquid that is denser than water and does not readily dissolved in water.

Discharge: The transfer of water out of an aquifer usually to a surface water body, a well or a spring.

Discharge Area: An area in which discharge occurs from an aquifer.

Divide (groundwater): An imaginary line which separates the areas within an aquifer in which groundwaters flow to different points of discharge. It is defined by the potentiometric or water table contours.

Divide (surface water): An imaginary line which separates surface water drainage basins in which surface runoff and streams flow to different surface water bodies. It is defined by the topographic expression of the land surface.

Drawdown: The decline in groundwater level at a point as a result of the withdrawal of water from an aquifer.

Downgradient: In the direction of decreasing groundwater head, in the direction of groundwater flow.

Driven Well: A shallow well installed by driving a pipe and screen into the aquifer.

Dug Well: A shallow well excavated by hand.

Effluent: The liquid discharge from a water treatment facility.

Eutrophication: A natural process (sometimes speeded up by unnatural processes) in which ponds and lakes become enriched with nutrients. This may result in weed growth and algae blooms.

Evaporation: The conversion of surface water to water vapor, usually through the sun's heat.

Evapotranspiration: The transfer of water from the surface to the air by a combination of evaporation and plant transpiration.

Fault: A fracture, or zone of fractures in a rock along which there was movement in the past.

FIFRA: The Federal Insecticide, Fungicide and Rodenticide Act.

Fracture: A linear opening in a rock.

Fracture Trace Analysis: An investigation aimed at determining the presence of regional fractures in rocks. This process usually involves examination of aerial photographs.

Gradient (groundwater or water table): The change in elevation of the water table or piezometric surface between two points divided by the distance between them. The slope of the water table or piezometric surface.

Gravel Pack: Well graded gravel or sand installed between a well screen and an unconsolidated formation in order to increase the efficiency and long term viability of the well.

Gravel Packed Well: A well installed in an unconsolidated deposit with a gravel pack around the well screen.

Groundwater: Underground water, usually the water in an aquifer below the water table.

Groundwater Basin: The area of an aquifer which contributes groundwater to a common point or area of discharge (such as a river or lake).

Groundwater Divide: An imaginary line which separates the areas within an aquifer in which groundwaters flow to different points of discharge. It is defined by the potentiometric or water table contours.

Hardness: Simply put, hardness is a measure of the ability of water to make bubbles in the presence of soap. Chemically, it is sometimes a measure of the concentration of calcium and magnesium in solution.

Hazardous Waste: Any of a number of chemicals, or mixtures of chemicals which: 1) are defined as hazardous by the EPA or state government, 2) which have been used in some manner and, 3) which have been disposed of or released to the environment.

Head: Usually, this term refers to the height of a column of water (such as the elevation of the water level in a well). Technically it is the potential energy in a mass of water produced by elevation or pressure.

Heavy Metals: Metals with a high molecular weight, most of which are dangerous to human health if ingested at certain concentrations.

Hydraulic Conductivity: A measure of the permeability of aquifer material. Defined as the rate of flow of water through a unit cross section of aquifer under a unit hydraulic head.

Hydraulic Gradient: The slope of a water table or potentiometric surface. The change in groundwater head over a given distance in a given direction.

Hydrogeology: The science of the occurrence and movement of groundwater in geologic formations.

Hydrogeologist: A scientist who studies the occurrence and movement of groundwater in geologic formations. The term implies that the scientist is trained as a geologist.

Hydrologist: A scientist who studies the occurrence and movement of surface water (usually streams) and sometimes groundwater.

Hydrology: The science of the occurrence and movement of water in natural systems. Hydrology is often thought of as the study of streams, but this branch of science includes all aspects of the hydrologic cycle, including groundwater.

Hydrologic Cycle: The cycle of water movement from atmosphere to land, streams lakes, aquifers, the sea and back again to the atmosphere.

Infiltrate: To seep into or through, especially soils.

Infiltration Capacity: The capacity of a soil or rock to take in water from precipitation or runoff.

Ion: A charged particle in solution.

Jetted Well: A well installed by "jetting water" to displace the soils.

Leachate: A solution of dissolved chemicals formed by water infiltrating through a landfill.

LUST: Leaking underground storage tank.

MCL: see Maximum Contaminant Level.

Maximum Contaminant Level: An EPA-defined concentration or level of contaminant in drinking water. Concentrations at or above the maximum contaminant levels are considered potentially hazardous to human health.

Maximum Contaminant Level Goals: An EPA-defined, but non-enforceable, concentration of a drinking water contaminant that is aimed to prevent adverse human health effects and allows an adequate margin of safety. These are goals, not enforceable levels.

Monitoring Well: A well designed or installed specifically for the purpose of providing groundwater samples for testing.

Naturally Developed Well: A well which is screened in unconsolidated deposits without a gravel pack.

Non-point Source: A widely spread, or diffuse source of contamination to groundwater such as road salts and nitrates from lawns and septic systems.

Numerical Model: A computer technique for predicting groundwater flow based on a finite difference or finite element grid. These types of models can be designed to simulate very complex aquifer systems and well pumping situations.

Observation Well: A well specifically designed or installed for the purpose of measuring water levels.

Outcrop: A rock formation or unconsolidated deposit that appears at the ground surface.

Overland Runoff: The portion of precipitation that flows overland or in streams once it has reached the earth's surface. Same as *Surface Runoff.*

Perched Groundwater: Unconfined groundwater which is separated from a lower, main body of groundwater by a confining or semi-confining bed.

Percolate: To infiltrate through.

Permeability: The capacity of a porous medium to transmit water. Same as **hydraulic conductivity.**

Piezometer: A non-pumping well, usually of small diameter, which is used to measure the elevation of the **water table** or **piezometric surface.**

Piezometric Level: The elevation of the water surface in a well or piezometer screened within a confined aquifer. Same as **potentiometric level**.

Piezometric Surface:. An imaginary surface connecting the potentiomentric levels of an aquifer. It is a general term which applies to both water table and confined aquifers, but is usually used in reference to confined aquifers. A water table is the potentiometric surface of an unconfined aquifer (with no vertical gradients). Same as **potentiometric surface.**

pH: A measure of the acidity or alkalinity of a solution. A pH of 7 is neutral. Acidity increases with lower pH, alkalinity increases with higher pH.

Phreatic Zone: The portion of an unconfined aquifer below the water table. Same as **saturated zone.**

Plume: A volume of contaminated groundwater in an aquifer.

Point Source: A single, identifiable, discreet source of contamination to groundwater. As opposed to a **non-point source.**

Pore Space: The openings in rocks and soils.

Porosity: The percentage of the volume of a body of rock or soil which is pore space and could be filled with water or air.

Potentiometric Level: The elevation of the water surface in a well or piezometer screened within a confined aquifer.

Potentiometric Surface: An imaginary surface connecting the potentiomentric levels of an aquifer. It is a general term which applies to both water table and confined aquifers, but is usually used in reference to confined aquifers. A water table is the potentiometric surface of an unconfined aquifer (with no vertical gradients). Same as piezometric surface.

Precipitation: The transfer of water from the atmosphere to the ground as rain, snow, hail or any other form.

Primary Openings: The open space between soil grains. Where water occurs in an unconsolidated aquifer.

Pump and Treat: A technique for attempting to treat or contain contaminated groundwater. Contaminated groundwater is pumped from a well or series of wells, piped through a treatment system and discharged, either back into the aquifer or to another discharge point (such as a surface water).

Pumping Test: A test conducted on a well to evaluate the performance of the well and of the aquifer and to calculate aquifer coefficients.

RCRA: The Resource Conservation and Recovery Act.

Recharge: 1) The transfer of water to an aquifer (usually through rainfall). 2) The water which is added to an aquifer.

Recharge Area: That portion of land which water flows over or under (ie., by way of runoff or groundwater flow) to reach a given area (e.g., a stream or lake) or point (e.g., a well or spring).

Remediation: The process of either: 1) treating contaminated soils or groundwater or 2) eliminating or minimizing the potential threat to human health and the environment posed by contaminated soils or groundwater.

Runoff: See surface runoff.

Saturated Zone: The portion of an unconfined aquifer below the water table.

Safe Yield: The maximum rate which water could be withdrawn from a well or spring without causing negative impacts such as excessive water level drawdown, aquifer dewatering or lowered water quality.

SDWA: The Safe Drinking Water Act.

Secondary Openings: Open space in rock caused by fracturing, solution or other means.

Seepage Velocity: The average rate of groundwater flow through a given portion of an aquifer.

Semi-confining Unit: In reality, all confining units are semi-confining in the sense that a small quantity of water can always leak through. The term semi-confining unit is sometimes used to describe a relatively leaky confining unit.

Septage: That which is pumped out of a septic system during cleaning.

Septage Lagoon: A trench, or series of trenches, into which septage is dumped.

Septic System: An on-site, underground system for disposing of the wastes which flow through plumbing. Typically consists of a septic tank and leaching field.

SOC: Synthetic Organic Compounds.

Soil: There are many possible technical definitions of **soil** but for the purposes of this book we will simply define it as the granular material formed from the decomposition of rock.

Soil Moisture: Water held temporarily in the soil.

Specific Conductivity: A measure of the ability of a liquid to conduct electricity. A common measure of water quality.

Specific Yield: A measure of the amount of water released from a volume of an aquifer as a result of a decline in head.

Storage Coefficient: The percent of water which is released when the water table or piezometric head drops in an aquifer.

Superfund: See **CERCLA**.

Surface Runoff: The portion of precipitation that flows overland or in streams once it has reached the earth's surface.

Surface Water: Any body of water on the surface of the land, such as rivers and lakes.

TCE: Trichloroethylene, a common **volatile organic compound**.

THMs: Trihalomethanes, volatile organic compounds sometimes formed when chlorine is added to waters with organic compounds.

Till: A relatively dense and poorly sorted deposit of sand, gravel, silt and sometimes clay which was deposited directly by glacial ice. Usually till is a very poor conductor of water.

Total Dissolved Solids: The total concentration of dissolved materials in water.

Transmissivity: A measure of the permeability of the entire thickness of an aquifer. Theoretically, the rate at which water would be transmitted through a unit width of an aquifer a unit hydraulic gradient. This is equivalent to the **hydraulic conductivity** multiplied by the saturated thickness of an aquifer.

Transpiration: The evaporation of water from plants and animals.

Tubular Wellfield: A group of small diameter wells (often 2.5 inch) connected together in a vacuum pumping system.

Unconfined Aquifer: An aquifer in which the groundwater is exposed to the atmospheric pressure through openings (pore space) in the overlying formation. Same as **water table aquifer.**

Unconsolidated Aquifer: An aquifer composed of unconsolidated materials such as sand and gravel, as opposed to consolidated materials such as **bedrock.**

Unsaturated Zone: The portion of a geologic formation above the water table of an unconfined aquifer.

Upgradient: In the direction of increasing groundwater head, in the opposite direction of groundwater flow.

USGS: United States Geological Survey.

USTs: Underground storage tanks.

Vadose Zone: The portion of a geologic formation above the water table of an unconfined aquifer. Same as **unsaturated zone.**

VOCs: See **volatile organic compounds.**

Volatile Organic Compounds: A group of organic chemicals which are common groundwater contaminants. This chemical group includes components of gasoline and other petroleum products as well as many common solvents. Many of these chemical are health hazards at very low concentrations.

Wastewater Treatment Facility: A general name for any facility designed to treat wastewater but most often refers to a large public facility for treating sewage.

Water Table: The surface of a body of unconfined groundwater at which the pressure is equal to that of the atmosphere.

Water Table Aquifer: An **unconfined aquifer.**

Water Table Contours: Lines on a map representing points of equal water table elevation. Used to determine groundwater flow direction.

Watershed: The area which contributes surface runoff to a common point or area (such as a river or lake). Same as a surface water **basin.**

Well Casing: The rigid, cylindrical portion of the well which provides support and integrity.

Wellhead: A water source, such as a well or spring.

Wellhead Protection Plan: A plan implemented for the purpose of protecting the water quality within the recharge area of a public water supply well. Protection measures might include land use zoning bylaws, underground storage tank regulations, land use restrictions and water quality monitoring.

Wellhead Protection Area: The area around a well in which a Wellhead Protection Plan is implemented. Usually based on some type of zone of contribution.

Well Screen: The slotted portion of a well which allows water to flow into the well but holds back the larger soil particles.

WHP: Wellhead Protection.

WHPA: Wellhead Protection Area.

Zone of Aeration: See **unsaturated zone**

Zone of Contribution: The land area which contributes recharge to a well under a specific set of circumstances. Often used to define a **Wellhead Protection Area**.

Zone of Saturation: The portion of an unconfined aquifer below the water table. Same as **saturated zone.**

References

Ainsworth, S., 1995, "Overview of Surface, Groundwater Quality," Opflow, Vol.21, No.3, American Water Works Association.

The American Society of Dowsers, 1980, *The Water Dowsers Manual*, The American Society of Dowsers, Danville, Vermont.

Back, W., J.S. Rosenshein, P.R. Seaber, 1899, *Hydrogeology*, The Geology of North America, Volume O-2, The Geological Society of America, Boulder, Colorado.

Bird, C., 1979, *The Divining Hand*, E. P. Dutton, New York.

Biswas, A.K., 1972, *History of Hydrology*, North-Holland Publishing Co., London.

Brackley, R.A. and B.P. Hansen, 1977, "Water Resources of the Nashua and Souhegan River Basins, Massachusetts," U.S. Geological Survey Hydrologic Investigations, HA-276.

California State University, School of Engineering, 1987, "Small Water System Operation and Maintenance," Office of Water Programs, California State University, Sacramento, 6000 J. Street, Sacramento, CA 95819-6025.

Cothern, C.R. and P.A. Roberts, 1991, *Radon, Radium and Uranium in Drinking Water*, Lewis Publishers, Chelsea, Michigan.

Darcy, H. 1856, *Les Fontaines Publiques de la Ville de Dijon*, V. Dalmont, Paris.

Driscoll, F.G., 1986, *Groundwater and Wells*, Johnson Division, UOP, Inc., St. Paul, Minn.

Fetter, C.W., 1992 , *Contaminant Hydrogeology*, Macmillan Publishing Company, New York.

Freeze, R.A. , and J.A. Cherry, 1979, *Groundwater*, Prentice Hall, Englewood Cliffs, N.J.

Gibb, J.P., 1973, "Planning a Domestic Water System," Illinois State Water Survey, Urbana, IL, ISWS-73-CIR116.

Haeni, F.P., 1995, "Applications of Surface-Geophysical Methods to Investigations of Sand and Gravel Aquifers in the Glaciated Northeastern United States," U.S. Geological Survey Professional Paper 1415-A.

Heath, R.C., 1987, "Basic Groundwater Hydrology," U.S. Geological Survey Water-Supply Paper 2220.

King, J., 1985, *Troubled Water*, Rodale Press, Emmaus, Pennsylvania.

Lehr, J.H., T.E. Gass, W.A. Pettyjohn and J. DeMarre, 1988, *Domestic Water Treatment*, National Water Well Association, Dublin, Ohio.

National Research Council, 1994, "Alternatives for Groundwater Cleanup."

Northeast Regional Agricultural Engineering Service, 1995, *Home Water Treatment*, Cooperative Extension, Ithaca, N.Y.

Pye, V.I., R. Patrick and J. Quarles, 1983, *Groundwater Contamination in the United States*, University of Pennsylvania Press, Philadelphia.

Schwalbaum, W.J., 1994, "A Numerical Groundwater Flow Model of the Martha's Vineyard Aquifer: Applications for Wellhead Protection and the Siting of a Wastewater Treatment Facility," Proceedings of the 1994 FOCUS Conference on Eastern Regional Ground Water Issues, National Ground Water Association, Dublin, Ohio.

Todd, D.K., 1980, *Groundwater Hydrology*, John Wiley and Sons, Inc., New York, N.Y.

U.S. Department of the Interior (Bureau of Reclamation), 1985, *Ground Water Manual*, U.S. Government Printing Office, Denver, Colorado.

U.S. Environmental Protection Agency, 1977, "The Report to Congress: Waste Disposal Practices and Their Effects on Groundwater," EPA 570/9-77-001, Washington, DC.

U.S. Environmental Protection Agency, 1982, "Manual of Individual Water Supply Systems," EPA 570/9-82-004, Center for Environmental Research Information, Cincinnati, Ohio.

U.S. Environmental Protection Agency, 1984, "Extent of the Hazardous Release Problem and Future Funding Needs," Office of Solid Waste and Emergency Response.

U.S. Environmental Protection Agency, 1985, "Protection of Public Water Supplies from Groundwater Contamination," EPA/625/4-85/016, Center for Environmental Research Information, Cincinnati, Ohio.

U.S. Environmental Protection Agency, 1987a, "Handbook: Ground Water," EPA/625/6-87/016, Center for Environmental Research Information, Cincinnati, Ohio.

U.S. Environmental Protection Agency, 1987b, "Guidelines for Delineation of Wellhead Protection Areas," Office of Groundwater Protection.

U.S. Environmental Protection Agency, 1990, "Ground Water, Volume 1: Ground Water and Contamination," EPA/625/6-90/016a, Center for Environmental Research Information, Cincinnati, Ohio.

U.S. Environmental Protection Agency, 1992a, "National Survey of Pesticides in Drinking Water Wells", Phase II Report (Chapter 6)

U.S. Environmental Protection Agency, 1992b, "National Water Quality Inventory," (Chapter 6).

U.S. Environmental Protection Agency, 1993, "Wellhead Protection: A Guide for Small Communities," EPA/625/R-93/002, Center for Environmental Research Information, Cincinnati, Ohio.

U.S. Environmental Protection Agency, 1994, "Is Your Drinking Water Safe?," EPA 810-F-94-002, Office of Water.

U.S. Environmental Protection Agency, 1996, "Drinking Water Regulations and Health Advisories," Office of Water.

Weimer, R.A., 1980, "Prevent Ground Water Contamination Before It's Too Late," Water and Wastes Engineering, February, pp 30-33.

INDEX

A

B

C